# BUILD YOUR OWN HOME
## on a Shoestring

Henry Nonnenberg Jr., Gwen Moran,
and Chris Bradley

ALPHA

A member of Penguin Group (USA) Inc.

*For Abigail, Owen, and Rhiannon*

**ALPHA BOOKS**

Published by the Penguin Group

Penguin Group (USA) Inc., 375 Hudson Street, New York, New York 10014, U.S.A.

Penguin Group (Canada), 10 Alcorn Avenue, Toronto, Ontario, Canada M4V 3B2 (a division of Pearson Penguin Canada Inc.)

Penguin Books Ltd, 80 Strand, London WC2R 0RL, England

Penguin Ireland, 25 St Stephen's Green, Dublin 2, Ireland (a division of Penguin Books Ltd)

Penguin Group (Australia), 250 Camberwell Road, Camberwell, Victoria 3124, Australia (a division of Pearson Australia Group Pty Ltd)

Penguin Books India Pvt Ltd, 11 Community Centre, Panchsheel Park, New Delhi—110 017, India

Penguin Group (NZ), cnr Airborne and Rosedale Roads, Albany, Auckland 1310, New Zealand (a division of Pearson New Zealand Ltd)

Penguin Books (South Africa) (Pty) Ltd, 24 Sturdee Avenue, Rosebank, Johannesburg 2196, South Africa

Penguin Books Ltd, Registered Offices: 80 Strand, London WC2R 0RL, England

International Standard Book Number: 1-59257-381-9
Library of Congress Catalog Card Number: 2005925419

07  06  05      8  7  6  5  4  3  2  1

Interpretation of the printing code: The rightmost number of the first series of numbers is the year of the book's printing; the rightmost number of the second series of numbers is the number of the book's printing. For example, a printing code of 05-1 shows that the first printing occurred in 2005.

*Printed in the United States of America*

**Note:** This publication contains the opinions and ideas of its authors. It is intended to provide helpful and informative material on the subject matter covered. It is sold with the understanding that the authors and publisher are not engaged in rendering professional services in the book. If the reader requires personal assistance or advice, a competent professional should be consulted.

The authors and publisher specifically disclaim any responsibility for any liability, loss, or risk, personal or otherwise, which is incurred as a consequence, directly or indirectly, of the use and application of any of the contents of this book.

Most Alpha books are available at special quantity discounts for bulk purchases for sales promotions, premiums, fund-raising, or educational use. Special books, or book excerpts, can also be created to fit specific needs.

For details, write: Special Markets, Alpha Books, 375 Hudson Street, New York, NY 10014.

**Publisher:** Marie Butler-Knight
**Product Manager:** Phil Kitchel
**Senior Managing Editor:** Jennifer Bowles
**Senior Acquisitions Editor:** Mike Sanders
**Development Editor:** Christy Wagner
**Senior Production Editor:** Billy Fields

**Copy Editor:** Keith Cline
**Cover Designer:** Ann Marie Deets
**Book Designer:** Trina Wurst
**Indexer:** Tonya Heard
**Layout:** Ayanna Lacey
**Proofreading:** Donna Martin

# CONTENTS

1   Let Us Help You Save Money on Your Dream Home ...............................................1

2   Financing Your Home .................................29

3   Choosing Your Location ............................53

4   Designing Your Dream Home ....................73

5   Saving on Your Structure...........................87

6   Finalizing Your Floor Plan......................101

7   Ready, Set, Build! ...................................125

8   All Systems Go.......................................155

9   Inside Jobs ............................................181

10  Kitchens, Baths, and Utility Rooms .........211

11  Landscaping, Driveways, and Exterior Living Spaces .............................................235

12  Contractors, Estimates, and Suppliers .....253

13  Estimating, Scheduling, and Budgeting Your Project ..........................................279

14  The Home Stretch ..................................299

## Appendixes

A   Glossary ................................................315

B   Resources .............................................325

C   Sample Documents ................................335

    Index ...................................................347

# INTRODUCTION

Wouldn't it be great to have a home that makes you say *"Ahhh"* every time you drive up to it? More people than ever are building their dream homes. What many of them don't know, however, is that a few tricks of the trade can save them thousands of dollars and as many headaches—and allow them to build a bigger or better house for their money.

Here's where you have the advantage. Whether your project is a luxurious estate home or a small cottage by a lake, *Build Your Own Home on a Shoestring* shares those money-saving tips, tricks, secrets, and ideas to help you save big bucks on your home's construction. From creating a floor plan to finding contractors to securing financing, you'll learn what the pros know—and what they don't want you to find out!

Best of all, this isn't a long, boring lecture. Instead, you'll find the information organized in a logical order and presented in an easy-to-read format. We know you're busy, so we've created a book that gives you just the facts in a fun and informative manner.

Our author team is sharing what we've learned by doing, as well as input and ideas from a number of professionals in every area of home construction. For nearly two decades, Chris Bradley has been a licensed general contractor based in Merrick, New York. Husband-and-wife team Hank Nonnenberg and Gwen Moran have been renovating and constructing their own homes since 1997, and Hank has been a licensed residential real estate appraiser for more than a decade. Together, they bring you an extensive pool of knowledge and will help you avoid common money traps and mistakes.

## Acknowledgments

Just like building a dream home, it takes many people to build a book. We'd like to thank some of the outstanding professionals who shared their expertise, including Barbara Reinecke of Pasch Realty (www.paschrealty.net); Bob Owen of Owen Plumbing and Heating (www.owenplumbinginc.com); Shaun Moran and Joe Low of Allied Capital Mortgage (www.alliedcapital.com); Patrick Egan of Hammers, Inc. (www.hammersinc.com); Dean Bennett of Dean Bennett Design and Construction (johnnydollarco@yahoo.com); Dan Bollin of Transtar Electric Security and Technology (www.transtarcorp.com); and Linda Reimer and Carl Cuozzo of Design Basics, Inc. (www. designbasics.com).

Thanks to our energetic and tireless agent, Marilyn Allen, as well as Mike Sanders, Christy Wagner, and the rest of the visionary crew at Alpha Books.

Chris thanks Tony Lentini, an amazing painter and wall-paper hanger, who took him under his wing and taught him the importance of craftsmanship, quality, and commitment, and his father-in-law, Steve Sachs, for teaching him all there is to know about the exterminating business. He also thanks his wife, Liz, for her support, for her business mind, and for telling him when he's wrong, as well as being a wonderful mother to Owen and Rhiannon.

Gwen and Hank thank their families, especially Lorraine Nonnenberg, Heather Nonnenberg, Jeanice Moran, and Tom Moran, who provided beyond-the-call-of-duty baby-sitting services during the writing of this book. And to Abigail, who provides the best inspiration to keep building the dream.

## Trademarks

All terms mentioned in this book that are known to be or are suspected of being trademarks or service marks have been appropriately capitalized. Alpha Books and Penguin Group (USA) Inc. cannot attest to the accuracy of this information. Use of a term in this book should not be regarded as affecting the validity of any trademark or service mark.

# 1

# LET US HELP YOU SAVE MONEY ON YOUR DREAM HOME

Welcome to *Build Your Own Home on a Shoestring*. We've written this book to share some of our firsthand lessons and knowledge about building a home. By doing so, we hope we help you avoid some mistakes and save a great deal of money when you're building your own dream home.

The chapters are arranged in sequence, following the order of decisions and actions you need to make and take in the construction of your new dwelling. Throughout the text, we included money-saving tips, ideas, and strategies to help you reap the same savings that professional contractors earn, while saving that 25 percent mark-up on services.

We suggest that you complete the worksheets and suggested actions (such as visiting model homes for ideas) as you read the book. After you've read it through, refer back to the book as a step-by-step guide to walk you through the process of building your home. Be sure to keep your own tally of savings as you incorporate the tips we share in these pages.

Keep in mind that this book does not provide legal advice. We are not lawyers and have no desire to be lawyers, so please don't construe any of our advice in that manner. In addition, there are many, many different climates, building codes, restrictions, and other factors that impact how you will choose, design, build, and finance your home, depending on where you live. So although we give tips and ideas that have worked for us and for many others, you should always double-check our suggestions with your contractor, building department, lender, or other relevant local authority before acting on them.

## Are You Ready?

It's woven into the fabric of the American dream as tightly as baseball and apple pie—building a dream home. Whether it's a small cottage by a lake, a 3-bedroom ranch in the suburbs, or a 5,000-square-foot estate in the country, the definition of "dream home" can vary widely from person to person and family to family.

Americans are building new homes at breakneck speed— 1.7 million of them in 2004, according to the National Association of Homebuilders. These homes are getting larger and more complex, on average, with more than 400 extra square feet of space over the past decade, and include 4 or more bedrooms, garages accommodating 2 or more cars, and more sophisticated amenities such as high-tech alarm systems, media rooms, central vacuum systems, and other features.

However, with these increases and extras usually comes an increased price. What can you do if you have big ideas for your dream home but only a small budget? You can incorporate simple money-saving ideas and strategies in the planning phase that continue to yield benefits long after the home has been built—and we show you how.

And although many people opt to choose from a predetermined style of home in new developments, you can actually save more money—and get a home that truly reflects your individual style, needs, and wants—by designing and building your own home. You can do this with a general contractor or act as your own general contractor.

## Contractor Basics

General contractors (GCs) are the people or companies that oversee all of the many functions that go into building a home, from excavating to framing to siding to plumbing to painting and everything in between. For the service of streamlining these functions for you, as their fee, they usually add 15 to 25 percent to the cost of each function.

If you're looking for a GC to help you build your home, you'll find a wide variety with an even wider variety of capabilities and services. Most GCs fall into the category of either custom homebuilders or production homebuilding companies. According to the National Association of Homebuilders, telling the difference can be tricky. It put together this quick rundown to show the difference between custom and production homebuilding companies:

*Custom homebuilders generally ...*

- Build on land you own. Some custom builders also build on land they own.
- Build custom, one-of-a-kind houses. These homes are site-specific homes built from a unique set of plans for specific clients. Some custom builders may offer design/build services.

- Build single-family homes, rather than apartments, condos, or commercial units.
- Are generally small-volume builders. (They build 25 or fewer homes a year.)
- Tend to build higher-end homes.

*Production homebuilders generally ...*

- Build on land they own.
- Tend to use stock plans, but usually offer a variety of plan choices and options.
- Build all types of housing—single-family, condos, town houses, and rental properties.
- Are large-volume builders. (They build more than 25 homes a year.)
- Generally build for all price points—entry level, move up, luxury, etc.

Unless you're having a home built in a pre-existing development, if you're going to use a general contractor, you're probably going to want to choose a custom home builder. Custom homebuilders tend to be more flexible in working with their customers, which may offer you more savings opportunities. For instance, you'll have more material and design options, which can significantly impact your final cost.

## On Your Own—or Almost

Another option you have is to act as your own general contractor. That means you hire all the subcontractors yourself and manage the overall construction process. You're responsible for hiring, scheduling, and paying each of the workers who will participate in building your home. By doing this, you can save the

25 percent the general contractor would get—that equals $50,000 on a $200,000 home!

Sound easy? Well, there are some challenges and some things you need to know before you get to work. In the upcoming chapters, we discuss the different options you have throughout the process of building your dream home. Often, the key to bringing your new residence in on-budget while still making it the home of your dreams is to make key compromises. By understanding your options and the cost differential and then making informed decisions about which options matter to you—and which options don't—you can save a significant amount of money in the construction of your home.

Finally, you can have one of your subcontractors, usually a framing contractor or carpenter, assist you with the project management. This individual will consult with you and may offer recommendations for other contractors, building methods, and the like. However, you only pay this individual for his or her time, and you save the percentage mark-up on services. So if you can't handle the whole thing yourself or feel like you need some guidance, this may be a cost-conscious solution.

## Defining Your Dream

Before you begin your building quest, you need to get a clear idea of what the phrase "dream home" means to you. Take a few minutes to think about the home that would best suit your lifestyle and family.

You have many options to consider, but chances are you already have an idea of what your dream home is. However, your dream will certainly become a nightmare if you don't ground it in the realities of your lifestyle and family situation. For instance, a small, lakeside cottage might seem like a great idea, but if you have five children, it could get a bit cramped.

Similarly, your country estate could become a major headache if it's located miles away from the mass-transit system you take to work.

### Reality Check

Be sure to think carefully about how your dream can be effectively incorporated into the reality of your world. Consider such things as how many people will be living in the home and what their bedroom and bathroom accommodations will need to be. Will your family grow or stay the same size? Do you need space for entertaining, or do you lead a more solitary lifestyle? Do any family members have special needs, such as disabilities that may require ramps and single-floor living, or severe asthma or allergies that may require you to forego rugs and opt for hot water baseboard heating systems? Paying close attention to how your dream home will accommodate your family's needs will ensure that it stays your dream home for years to come.

### Types of Houses

Houses come in many different types and styles. Here we've listed the most common types and their identifying features:

- **Ranch.** This style of house is a single-story structure that is rectangular, L-shaped, or U-shaped. The roof is usually low pitched, and the home usually has an attached garage.
- **Cape Cod.** Traditionally, this style of home is 1½ stories, meaning that only a portion of the second floor is accessible for living space. Cape Cods have steep, gabled roofs and dormers. The facade is usually symmetrical, and the door is located in the center of the house.
- **Colonial.** Colonials are usually 2-story homes. The first floor is typically the living area, such as the kitchen, living room, and the like. The bedrooms are usually located on

the second floor. However, some Colonials might have a bedroom or suite located on the first floor.

- **Bi-level.** This type of home has an entrance to a foyer from which you can walk upstairs or downstairs. The upstairs typically has a kitchen and living area as well as bedrooms. The downstairs may have living areas or bedrooms.

- **Split level.** Split-level homes have multiple living levels with small stairways leading to each level. They are typically asymmetrical, and the entrance is on the bottom floor.

- **Raised ranch.** This style of house features a basement area that is partially above ground. The front door typically opens to the second floor, with stairs leading down to the living areas. The upstairs area houses the kitchen and bedrooms.

- **Modern.** Modern homes have sleek lines and unusual angles. They may be symmetrical or asymmetrical, and they may be single or multiple stories.

You have many types of homes from which to choose. Determine what your essentials are and then choose or develop a plan that best meets those needs. For instance, if you have aging family members or individuals with disabilities living in your home, it may be best to choose a single-story home or a home with living quarters on the first floor. If your family is growing, you may want to choose a home style that has extra bedrooms or room to expand so you have the additional space when you need it.

# GO FIGURE

Use this checklist to determine the most important features of your dream home. Check Must Have, Nice to Have, or Not Important to indicate the priorities for your home.

Number of bedrooms: _____

Number of bathrooms: _____

Style of home (Ranch, Colonial, Cape Cod, etc.) _____

Other important features:

_____

_____

**Home Features:**

| | Must Have | Nice to Have | Not Important |
|---|---|---|---|
| *Kitchen:* | | | |
| Eat-in kitchen | ☐ | ☐ | ☐ |
| No eat-in kitchen, dining room only | ☐ | ☐ | ☐ |
| Dishwasher | ☐ | ☐ | ☐ |
| Extra-large refrigerator | ☐ | ☐ | ☐ |
| Double ovens | ☐ | ☐ | ☐ |
| Wall oven | ☐ | ☐ | ☐ |
| Oven with cooktop | ☐ | ☐ | ☐ |
| Island | ☐ | ☐ | ☐ |
| Sink style: | | | |
| Single | ☐ | ☐ | ☐ |
| Double | ☐ | ☐ | ☐ |
| Ample counter space | ☐ | ☐ | ☐ |
| Flooring: | | | |
| Ceramic tile | ☐ | ☐ | ☐ |
| Hardwood | ☐ | ☐ | ☐ |
| Laminate | ☐ | ☐ | ☐ |
| Vinyl | ☐ | ☐ | ☐ |
| Other features: _____ | ☐ | ☐ | ☐ |
| _____ | ☐ | ☐ | ☐ |
| _____ | ☐ | ☐ | ☐ |

|  | Must Have | Nice to Have | Not Important |
|---|:---:|:---:|:---:|
| *Living Room:* | | | |
| Attached to dining room | ☐ | ☐ | ☐ |
| Fireplace | ☐ | ☐ | ☐ |
| No formal living room, family room only | ☐ | ☐ | ☐ |
| Built-in shelving or other units | ☐ | ☐ | ☐ |
| Other features: _____ | ☐ | ☐ | ☐ |
| _____ | ☐ | ☐ | ☐ |
| _____ | ☐ | ☐ | ☐ |
| *Dining Room:* | | | |
| Attached to living room | ☐ | ☐ | ☐ |
| Access to kitchen | ☐ | ☐ | ☐ |
| Eat-in kitchen, no dining room | ☐ | ☐ | ☐ |
| Other features: _____ | ☐ | ☐ | ☐ |
| _____ | ☐ | ☐ | ☐ |
| _____ | ☐ | ☐ | ☐ |
| *Family Room:* | | | |
| Located on first floor | ☐ | ☐ | ☐ |
| Fireplace | ☐ | ☐ | ☐ |
| Open access to kitchen | ☐ | ☐ | ☐ |
| Access to backyard or deck | ☐ | ☐ | ☐ |
| Built-ins (shelves, desk, etc.) | ☐ | ☐ | ☐ |
| Other features: _____ | ☐ | ☐ | ☐ |
| _____ | ☐ | ☐ | ☐ |
| _____ | ☐ | ☐ | ☐ |
| *Master Bedroom:* | | | |
| First floor | ☐ | ☐ | ☐ |
| Upper floor | ☐ | ☐ | ☐ |
| Bathroom attached | ☐ | ☐ | ☐ |
| Sitting area | ☐ | ☐ | ☐ |
| Fireplace | ☐ | ☐ | ☐ |
| Vaulted ceilings | ☐ | ☐ | ☐ |
| Walk-in closet | ☐ | ☐ | ☐ |
| Master bathroom | ☐ | ☐ | ☐ |
| Other features: _____ | ☐ | ☐ | ☐ |
| _____ | ☐ | ☐ | ☐ |
| _____ | ☐ | ☐ | ☐ |

|  | Must Have | Nice to Have | Not Important |
|---|---|---|---|
| *Family Bedrooms:* | | | |
| First floor | ☐ | ☐ | ☐ |
| Upper floor | ☐ | ☐ | ☐ |
| Bathroom attached | ☐ | ☐ | ☐ |
| Walk-in closet | ☐ | ☐ | ☐ |
| Other features: _____ | ☐ | ☐ | ☐ |
| _____ | ☐ | ☐ | ☐ |
| _____ | ☐ | ☐ | ☐ |
| | | | |
| *Master Bathroom:* | | | |
| Full bath | ☐ | ☐ | ☐ |
| Type of bathtub: | | | |
| Shower | ☐ | ☐ | ☐ |
| Whirlpool | ☐ | ☐ | ☐ |
| Steel or cast iron | ☐ | ☐ | ☐ |
| Fiberglass | ☐ | ☐ | ☐ |
| Separate shower | ☐ | ☐ | ☐ |
| Shower walls: | | | |
| Tile | ☐ | ☐ | ☐ |
| Tub surround | ☐ | ☐ | ☐ |
| Toilet separate from rest of bathroom | ☐ | ☐ | ☐ |
| Bidet | ☐ | ☐ | ☐ |
| Sink: | | | |
| Pedestal | ☐ | ☐ | ☐ |
| Cabinet | ☐ | ☐ | ☐ |
| Venting: | | | |
| Window | ☐ | ☐ | ☐ |
| Fan | ☐ | ☐ | ☐ |
| Other features: _____ | ☐ | ☐ | ☐ |
| _____ | ☐ | ☐ | ☐ |
| _____ | ☐ | ☐ | ☐ |
| | | | |
| *Additional Bathrooms:* | | | |
| Full bath | ☐ | ☐ | ☐ |
| Type of bathtub: | | | |
| Shower | ☐ | ☐ | ☐ |
| Whirlpool | ☐ | ☐ | ☐ |

|  | Must Have | Nice to Have | Not Important |
|---|---|---|---|
| Steel or cast iron | ☐ | ☐ | ☐ |
| Fiberglass | ☐ | ☐ | ☐ |
| Separate shower | ☐ | ☐ | ☐ |
| Shower walls: |  |  |  |
| Tile | ☐ | ☐ | ☐ |
| Tub surround | ☐ | ☐ | ☐ |
| Toilet separate from rest of bathroom | ☐ | ☐ | ☐ |
| Bidet | ☐ | ☐ | ☐ |
| Sink: |  |  |  |
| Pedestal | ☐ | ☐ | ☐ |
| Cabinet | ☐ | ☐ | ☐ |
| Venting: |  |  |  |
| Window | ☐ | ☐ | ☐ |
| Fan | ☐ | ☐ | ☐ |
| Other features: _____ | ☐ | ☐ | ☐ |
| _____ | ☐ | ☐ | ☐ |
| _____ | ☐ | ☐ | ☐ |

Features for additional bathrooms:

_____

_____

*Great Room/Library (typically on first floor):*

| | | | |
|---|---|---|---|
| Located in front section of house | ☐ | ☐ | ☐ |
| Located in rear section of house | ☐ | ☐ | ☐ |
| Closet | ☐ | ☐ | ☐ |
| Bathroom attached | ☐ | ☐ | ☐ |
| Built-ins | ☐ | ☐ | ☐ |

*Garage:*

| | | | |
|---|---|---|---|
| One-car | ☐ | ☐ | ☐ |
| Two-car | ☐ | ☐ | ☐ |
| Three-car | ☐ | ☐ | ☐ |
| Workbench | ☐ | ☐ | ☐ |
| Cabinets | ☐ | ☐ | ☐ |
| Front entrance | ☐ | ☐ | ☐ |
| Side entrance | ☐ | ☐ | ☐ |

| | Must Have | Nice to Have | Not Important |
|---|---|---|---|
| Attached | ☐ | ☐ | ☐ |
| Access from garage to home | ☐ | ☐ | ☐ |
| Other features: _____ | ☐ | ☐ | ☐ |
| _____ | ☐ | ☐ | ☐ |
| _____ | ☐ | ☐ | ☐ |

*Additional Rooms and Living Space:*

| | Must Have | Nice to Have | Not Important |
|---|---|---|---|
| Sun room | ☐ | ☐ | ☐ |
| Screened porch | ☐ | ☐ | ☐ |
| Screened patio | ☐ | ☐ | ☐ |
| Library/study | ☐ | ☐ | ☐ |
| Home office | ☐ | ☐ | ☐ |
| Workshop | ☐ | ☐ | ☐ |
| Children's playroom | ☐ | ☐ | ☐ |
| Gym/exercise room | ☐ | ☐ | ☐ |
| Sauna | ☐ | ☐ | ☐ |

*Exterior:*

| | Must Have | Nice to Have | Not Important |
|---|---|---|---|
| Siding | ☐ | ☐ | ☐ |
| Vinyl | ☐ | ☐ | ☐ |
| Full brick | ☐ | ☐ | ☐ |
| Accent brick | ☐ | ☐ | ☐ |
| Brick or stone veneer | ☐ | ☐ | ☐ |
| Full stone | ☐ | ☐ | ☐ |
| Accent stone | ☐ | ☐ | ☐ |
| Stucco | ☐ | ☐ | ☐ |
| Fiber cement | ☐ | ☐ | ☐ |
| Wood clapboard | ☐ | ☐ | ☐ |
| Wood shingle | ☐ | ☐ | ☐ |
| Cedar shake | ☐ | ☐ | ☐ |
| Other features: _____ | ☐ | ☐ | ☐ |
| _____ | ☐ | ☐ | ☐ |
| _____ | ☐ | ☐ | ☐ |

Roof:

| | Must Have | Nice to Have | Not Important |
|---|---|---|---|
| Asphalt shingles | ☐ | ☐ | ☐ |
| Wood shingles | ☐ | ☐ | ☐ |
| Slate | ☐ | ☐ | ☐ |
| Tile | ☐ | ☐ | ☐ |
| Synthetic materials | ☐ | ☐ | ☐ |

| | Must Have | Nice to Have | Not Important |
|---|---|---|---|
| Shingles: | | | |
| 3-tab | ☐ | ☐ | ☐ |
| Dimensional | ☐ | ☐ | ☐ |
| Synthetic | ☐ | ☐ | ☐ |
| Windows: | | | |
| Vinyl | ☐ | ☐ | ☐ |
| Wood/vinyl combination | ☐ | ☐ | ☐ |
| Wood | ☐ | ☐ | ☐ |
| Insulated glass | ☐ | ☐ | ☐ |
| Low-emissivity glass | ☐ | ☐ | ☐ |
| Upgraded hardware | ☐ | ☐ | ☐ |
| | | | |
| *Heating/Cooling:* | | | |
| Electric heat | ☐ | ☐ | ☐ |
| Oil heat | ☐ | ☐ | ☐ |
| Radiant heat | ☐ | ☐ | ☐ |
| Natural gas heat | ☐ | ☐ | ☐ |
| Electric baseboard system | ☐ | ☐ | ☐ |
| Forced hot air system | ☐ | ☐ | ☐ |
| Geothermal system | ☐ | ☐ | ☐ |
| Hot water baseboard system | ☐ | ☐ | ☐ |
| Solar heating system | ☐ | ☐ | ☐ |
| Central air conditioning | ☐ | ☐ | ☐ |
| Zoned systems (allows you to control the temperature in different areas of the home) | ☐ | ☐ | ☐ |
| Water softener | ☐ | ☐ | ☐ |
| Water heater | ☐ | ☐ | ☐ |
| | | | |
| *Utility Room:* | | | |
| First floor | ☐ | ☐ | ☐ |
| Second floor | ☐ | ☐ | ☐ |
| Gas dryer | ☐ | ☐ | ☐ |
| Electric appliances | ☐ | ☐ | ☐ |
| Sink | ☐ | ☐ | ☐ |
| Built-in cabinets | ☐ | ☐ | ☐ |
| Shelving | ☐ | ☐ | ☐ |
| Laundry chute (from upper floors) | ☐ | ☐ | ☐ |
| Other features: _____ | ☐ | ☐ | ☐ |
| _____ | ☐ | ☐ | ☐ |
| _____ | ☐ | ☐ | ☐ |

| | Must Have | Nice to Have | Not Important |
|---|---|---|---|
| *Foundation:* | | | |
| Slab or crawl space | ☐ | ☐ | ☐ |
| Finished for family use | ☐ | ☐ | ☐ |
| Unfinished for storage or utility use | ☐ | ☐ | ☐ |
| Partial basement | ☐ | ☐ | ☐ |
| Other features: _____ | ☐ | ☐ | ☐ |
| _____ | ☐ | ☐ | ☐ |
| _____ | ☐ | ☐ | ☐ |
| *Additional Needs:* | | | |
| Room or accommodations for pets | ☐ | ☐ | ☐ |
| Energy efficiency | ☐ | ☐ | ☐ |
| Extra-high ceilings | ☐ | ☐ | ☐ |
| Other features: _____ | ☐ | ☐ | ☐ |
| _____ | ☐ | ☐ | ☐ |
| _____ | ☐ | ☐ | ☐ |
| *Location:* | | | |
| Urban area | ☐ | ☐ | ☐ |
| Suburban area | ☐ | ☐ | ☐ |
| Rural area | ☐ | ☐ | ☐ |
| Proximity to mass transit | ☐ | ☐ | ☐ |
| Proximity to services and amenities (hospital, doctor, dry cleaner, etc.) | ☐ | ☐ | ☐ |
| Availability of public water | ☐ | ☐ | ☐ |
| Availability of public sewer | ☐ | ☐ | ☐ |
| Availability of natural gas | ☐ | ☐ | ☐ |
| Other features: _____ | ☐ | ☐ | ☐ |
| _____ | ☐ | ☐ | ☐ |
| _____ | ☐ | ☐ | ☐ |

## Build—or Rebuild?

If you currently own a home, you might be able to use many of the ideas in this book to remodel, add on to, or completely refurbish your current house to transform it into your dream home.

According to the American Homeowner Foundation, selling your home and moving typically costs about 8 to 10 percent of the value of your current home, including moving expenses, closing costs, and broker commissions. That's an expense beyond the costs of building a new home. So if your current residence is close to satisfying your needs, consider whether it's worthwhile to upgrade it.

## Finance and Housing Options to Consider

Whether you rent an apartment or have a residence you need to sell, it's important to have a game plan in place to deal with your existing financial commitments both while your new home is being built and before you move in. Because financing payments will be due before your home is completed, it's likely that you'll have payments due on your rent or mortgage as well as on your construction financing, so consider that added expense when you're figuring out your project finances.

Also be sure to have a backup housing plan. If you have a deadline by which you need to vacate your current residence, be sure you have options, just in case you meet unexpected construction delays. Ask your landlord if he or she will allow you to rent month-to-month after your lease is up, make arrangements for a short-term rental, or cash in some favors from family and friends who might let you stay with them for a while if your home construction takes longer than planned.

## Options, Options

It's no secret that building a new home is a big undertaking. You will have to manage literally hundreds of details—from finding and hiring people to work on your home to arranging for inspections to securing financing to scheduling subcontractors to choosing every element of the paint, floor coverings, moldings, tile, and other interior décor. Many people who purchase new homes do so from a model—where many of these decisions have been narrowed down to a few options—but building your dream home on your own puts total control of the final product in your hands. And although that means that you'll be able to choose exactly what you want from the myriad options available from various manufacturers, it also means you need to spend the time to do your homework, thinking about these decisions and the impact they will have on your home.

How can you find out about the options available to you? Home décor magazines, also called shelter publications, are great resources, showcasing designer applications of all the latest and greatest products. You'll find many titles on newsstand shelves. Some of the most popular are *House Beautiful, Better Homes and Gardens, Traditional Home, Metropolitan Home, Country Living, Elle Décor,* and *Architectural Digest.* Visit model homes in new developments in your area, which are often professionally designed, and ask about the products they used. Talk to the professionals at your local hardware and home design stores for free advice. Or check out the websites of your favorite brands of products to see their new offerings.

## Getting Into a Building Frame of Mind

Preparing for your building adventure is critical. It's exciting, and the end result will be a dream come true, but you need to understand that, like any worthwhile adventure, there will be challenges, headaches, and obstacles—as well as triumphs— along the way. Being involved and organized can help you avoid some of these challenges, but some will be out of your control. It's important to remain calm and deal with these challenges as they arise, and allow enough time in your scheduling to accommodate unforeseen bumps in the road.

The following sections offer some suggestions to help you make the building process less painful—and less expensive.

### Find Time

If you have a job that requires long days or won't allow you to accommodate occasional homebuilding demands during business hours, the process is going to be very stressful and will likely take longer than it otherwise could. Having the flexibility to meet with contractors and inspect their work during the day can significantly speed the building progress and eliminate potential misunderstandings. For instance, Hank and Gwen visited their job site on the day the excavators were beginning to dig the foundation, only to find that the area to be dug was marked incorrectly. Had the foundation been dug as marked, it would have needed to be corrected, adding time and expense to the project. Be sure you'll have the time you need to check on your home's progress.

### Get Organized

With so many decisions to make, contractors to hire, and projects to oversee, it's critical to keep track of your commitments, appointments, budgets, and suppliers. This can be as simple as creating a notebook or binder with sections for various elements of the project—excavating, masonry, framing, kitchen, bathrooms, wall and floor coverings, etc. A three-ring binder works well because you can fill it with clear plastic page protectors to hold magazine clippings or photos of designs you like for easy reference. For the more technically inclined, you can use your computer, laptop, or PDA to scan and hold files, schedule appointments, etc.

### Talk to Others

Find other folks who have been through the homebuilding process and get their advice. Visit online home-remodeling and homebuilding message boards, such as www.doityourself.com, www.homebuilding.about.com, www.taunton.com/finehomebuilding, or www.aecdaily.com. Read past messages to find out what pitfalls to avoid and ask questions if you have them.

Getting advice and firsthand tips from people who have lived through the experience of building their homes can significantly shorten your learning curve, provide you with tips and answers, and give you the benefit of the research that others have done and the resources they've uncovered. Hank and Gwen hired the same framing contractor used by a friend when he built his house. Even though they shopped around to compare prices, they decided to hire the contractor who came with a personal recommendation—and who gave a more competitive bid because he was referred by another customer.

### Take a Class

Check your local community college or building-supply store. Both usually have some sort of classes or seminars on various do-it-yourself topics. You could also check out a few home-building schools and courses online, such as those offered by HomeBuildingManual.com or Southern Home Building Seminars (www.southernhomebuildingseminars.com).

### Watch TV

Check out some of the homebuilding shows on network and cable television. Shows such as *Hometime, Bob Vila's Home Again,* and even *Trading Spaces* may give you ideas for your project—and insight about how to get it done.

### Spend Wisely

Don't spend more than a few dollars on anything before you do your homework and figure out what you can afford. Falling in love with home plans or spending money on an architect is

usually a bad idea—and a waste of money—until you know
you're ready to build.

## Preventing the Unexpected

The process of building your home can be tremendously gratify-
ing, but it can also be tremendously frustrating. Before you begin
the building process, you should really know what to expect and
try to anticipate what can go wrong, because it just might.

The following sections list some of the most common experiences of do-it-yourself general contractors.

### Plan for Delays

Whether it's an unexpected storm that pushes back your framing
schedule, an inspection that didn't go as planned, or a contractor who is running behind, delays will happen, so build time
into your schedule. Building deadlines into contractor agreements can be helpful, and planning your construction around
the weather for your area helps minimize delays as well.

### Budget Challenges

A devastating hurricane struck Florida just weeks before Hank
and Gwen placed a big lumber order for their dream home.
Because of increased demand, lumber prices skyrocketed, adding
about $10,000 to their projected lumber budget.

This type of budget overage happens often, but you can
usually balance your budget by cutting back on other expenses.
In our case, we chose a less-expensive window option to make
up most of the cost. Keep a close eye on your budget to know
where you have room to maneuver in case unexpected events
cause price increases.

### Happy—and Unhappy—Surprises

Just when you think you have everything under control and know your home inside and out, something will happen to change your perspective. For instance, one home-building team didn't realize that their plan called for a sunken family room. The feature didn't add to the cost of the construction, and ended up being a feature that the couple loves. Had they not wanted a sunken feature, however, it would have cost money to rework the plans. Pay attention to the myriad details on your floor plan. You might also consider finding someone who has experience reading plans to explain some of the mysterious symbols and point out any potential areas of concern. If you don't have a contact that immediately comes to mind, we'll discuss how to find a professional in Chapter 6.

### A Healthy Dose of Skepticism

You might go to great lengths to hire contractors whom you trust, but even then, it's important to check and double-check the work they're doing. Don't just assume the contractor is infallible. Instead, keep reviewing the plans and ensuring that the work being done matches the plans you have.

It's inevitable that you'll feel stressed through this process, but you should also try to enjoy it as much as possible. There are few feelings like watching your dream home take shape before your very eyes. Have your contingency plans in place (more about those later), remain flexible, and take a deep breath. You're in for an exhilarating ride!

## Fixing Your Finances

This is also the point when you should start getting your finances in order so you can qualify for the best possible financing. If you don't already, it's important to check your credit report at least once a year. At the very least, you should get copies of your credit report at least 6 to 8 months before you apply for financing. That way, if errors exist or if you have a history of late payments, which may lower your credit rating, you have time to take action to correct or improve your report.

## Credit Where Credit Is Due

Three primary companies compile the information lenders and other creditors use to determine whether to lend you money or extend credit, how much will be made available to you, and at what interest rate:

### Equifax
PO Box 740241
Atlanta, GA 30374-0241
Order credit report: 1-800-685-1111
Report fraud: 1-800-525-6285 and write to address above.

### Experian (formerly TRW)
PO Box 1017
Allen, TX 75013
Order credit report: 1-800-682-7654 or 1-888-397-3742
Report fraud: 1-800-301-7195 and write to address above. Look at your credit report for free at www.freecreditreport.com.

### TransUnion
PO Box 390
Springfield, PA 19064
Order credit report: 1-800-916-8800
Report fraud: 1-800-680-7289 and write to:

> Fraud Victim Assistance Division
> PO Box 6790
> Fullerton, CA 92634

Probably the most important part of your credit report is actually a credit score assigned to you; this might also be called a Fair Isaac and Company (FICO) score (named for the company that developed the scoring system). These scores generally range from 300 to 850 and take into consideration such factors as your past payment history, the amount of debt you owe, the length of your credit history, whether you've taken on new debt recently, any past judgments and bankruptcies, and the types of credit extended to you. If your credit report has errors or old information, this can negatively affect your ability to get a mortgage, reduce the amount for which you qualify, and increase your interest rate. It's very important to ensure that your credit report is as accurate as possible.

It's also important to obtain a copy of your report from all three credit bureaus, because errors on one report might not show up on all three. Clerical mistakes, mistaken identity, or other factors might have landed someone else's negative information on your report, so you should be sure that all three reports are as clean as possible.

### Evaluating Your Credit Report

Different lenders have different cutoffs for how they rank credit scores, but in general, this is how they rate.

| | |
|---|---|
| **720 and above** | *Excellent.* You'll have the most flexibility in the size of your loan and the best rates. |
| **700 to 719** | *Very good.* Again, you'll qualify for more money and better rates. |
| **680 to 699** | *Good.* You shouldn't have a problem finding a lender, and your rates should be competitive. |

**DON'T TRIP** on your **SHOESTRINGS**

A number of credit bureaus offer aggregate reports based on information from all three agencies. However, correcting information through these bureaus may not ensure that it's corrected on the reports issued by Experian, Equifax, or TransUnion, and the errors may still be on the report your lender checks. Go directly to the three major sources to have any errors on your credit report corrected.

620 to 679          *Adequate.* You should still be able to
                    find a lender, but you may require
                    a larger down payment or the loan
                    may cost you more.

If your score is below 620, you might be considered a greater risk, and your ability to negotiate the cost of the loan—points, interest rate, etc.—may decrease. Don't despair, however. Many lenders specialize in so-called "sub-prime" markets.

If you do have some rough spots on your credit history, such as late payments, overdue medical bills, and the like, it's important to keep on financial track in the months ahead. Even 6 months of on-time payments can significantly increase your credit score and lower the interest rate for which you'll qualify.

## Increasing Your Score

If your score is low because it contains errors, you have recourse. By law, you can challenge information on your report, and the credit bureau has 30 days to either verify the information or remove it. Contact the bureau in writing with your challenge. Do the same if the report contains information that should have been removed due to age. Negative credit information generally stays on the report for 7 years. Bankruptcies stay on for 10 years.

You can also increase your score by reducing new charges and paying off what you owe to reduce existing balances. After you pay off a card, it may not be a good idea to close it right away; FICO scores are based in part on the percentage of credit you have available to you versus the percentage of debt you have. So if you can be disciplined and not run up new charges, keep the card open and maintain a zero balance. If it's more likely that you'll end up charging a new balance, however, go ahead and close the card, because the increased debt may lower your score.

**DON'T TRIP**
on your **SHOESTRINGS**

Beware of quick credit fixes. Companies advertising that they can help you repair your credit quickly are often a scam at best and illegal at worst. These companies use myriad tricks, such as flooding the credit reporting bureau with bogus letters. Some create a new tax identification number through the IRS, which is illegal. If the consumer is caught, he or she can face stiff fines. If you need help cleaning up your credit, consult a reputable not-for-profit credit bureau. Find one through the National Foundation for Credit Counseling (www.nfcc.org).

## Qualifying for a Loan

How do you put your best foot forward to get the best financing at the best rates? Your credit scores are a big part of that equation, and if you need to clean them up, refer to the preceding sections. In addition, you can take other actions to show the lender that you're a good risk.

Overall, lenders are looking for three things:

- Your willingness to repay the debt
- Your ability to repay the debt
- The asset being financed

Your previous credit history is one component and shows your willingness to repay debt. However, lenders generally look for two out of the three components; so if you have a shaky credit history, you still have a chance to get the financing you need.

### Your Ability to Repay Debt

This includes a steady income, preferably showing a track record at the same job for at least 2 years. Lenders also look at the ratio of your housing expense (your mortgage payment, including principal and interest, property taxes, and insurance) to your overall monthly income. Lenders generally like to see that your housing expense is less than one third of your income—and the lower the ratio, the better. It's also best if your total monthly debts are no greater than 40 percent of your monthly income.

### The Asset Being Financed

Lenders want to see that the home you're building will have equity in it after it's built. They want to see that when the house is finished, it will be worth more than the mortgage amount you'll owe on it. The greater the amount of equity, the lower an investment risk it is for the lender, who will be able to secure his investment through the value of the property.

After you've chosen a finance plan, you can visit the tax assessor's office of the town in which you're going to build. Ask how you can access the SR-1A documents of recent sales. Look for the sale prices of homes in the area that are similar in size and specs (number of bedrooms, number of bathrooms) to what you're building.

You can also visit a Realtor, especially if you're using one to sell the home in which you currently live or will use one to purchase the land on which you will build. Ask for the sale prices of comparable properties. Your lender might also be able to access this information for you. Of course, your lender will also have a licensed real estate appraiser value your home after it's built, but having this information can help you better understand the amount for which you may qualify.

## GO FIGURE

Want to figure out your ratios? Use these worksheets:

**Housing Expense Ratio**

Your gross monthly income _____

Estimated mortgage payment (principal
 and interest) _____

Estimated property taxes (monthly) _____

Private mortgage insurance (for
 financing, which will be more than
 80 percent of the property's value) _____

**Total:** _____

÷ gross monthly income (GMI) _____

Percentage of GMI _____

**Debt-to-Income Ratio**

Your gross monthly income _____

Monthly credit card payments _____

Monthly car payments _____

Monthly student loan payments _____

Monthly revolving loan payments _____

Other monthly credit line payments _____

**Total monthly payments** _____

÷ GMI _____

Percentage of GMI _____

If your percentages are higher than the guidelines, keep in mind that you still may be able to qualify for financing, depending on your credit scores and the value of the property. Consult with your lender, or work to improve the ratios by cutting back your housing budget or decreasing your monthly debt before you apply for financing.

In the next chapter, we walk you through some of the critical things you need to know about financing your building process.

Suppose you qualify for a loan of about $100,000 more than you anticipated. Time to upgrade the game plan, right?

Hold on a minute. No lender knows your lifestyle like you. If you have good credit and a steady job, you may qualify for a much bigger loan than what you can comfortably take on. Consider the expenses you have that aren't reflected on your credit report—do you like to vacation or eat out frequently? What are some of your noncredit monthly expenses, such as commuting, groceries, entertainment, etc.? Do you have frequent medical expenses? Do you spend money on theater tickets? Do you have a hobby that costs money?

The discretionary income you have available to enjoy these things might dry up if you have a loan that's too high. Therefore, you need to consider what you can comfortably spend on your housing cost and be sure to leave yourself enough financial room to enjoy your life.

## Dollar-Saving Do's and Don'ts

- If you have the time and flexibility to be your own general contractor, you can save as much as 25 percent of the cost of your home.

- Get money-saving and project-management ideas by speaking with people who've been through the house-building process, reading home design and construction magazines or websites, and watching home-remodeling shows. Of course, reading this book is a great place to start, too.

- Don't spend a penny until you've done your homework.

- Figure out your monthly expenses so you know what kind of mortgage payment you can really afford.

- Get your credit in top shape to help reduce fees and interest on your loan.

# 2

# FINANCING YOUR HOME

The first stop in your hunt for the perfect home design isn't an architect, a real estate agent, or even a home design book. Before you spend a lot of time looking at floor plans and getting estimates, you should speak to a finance professional about what you can afford and the type of financing and terms for which you qualify. By knowing your financial parameters, you can begin to create a budget that makes sense for you. Then you can start making decisions about the type of home you're going to create.

## What You Can Expect When Applying for Your Loan

The process of applying for a construction loan or mortgage can be a nail-biting one. It seems like you have to answer hundreds of questions and provide your life history to be judged by some underwriter you've never even met.

Applying for a loan can be a bit labor-intensive, but it really consists of four parts:

- Determining the amount for which you can qualify
- Finding a lender
- Providing the paperwork
- Negotiating the terms

Of course, there's also the waiting; but if you've gotten your credit score in good shape and follow the advice in this chapter, you shouldn't have to wait for very long before you get access to the money you need.

## Show Me the Money!

By getting prequalified for a loan, you'll know roughly how much you can spend on building your home. Prequalification is basically applying for the loan before you go searching for your property. Based on your application, lenders can give you an idea of how much financing they can make available to you and at what terms. Of course, final approval of the loan then depends on the property being worth as much as or more than the amount you offer the seller, but having a prequalification means that you know the financing is approved, which gives you more bargaining power—sellers will take you more seriously because they know you can afford what you're offering. Getting prequalified can also shorten the length of time it takes to close the loan—another benefit that may make a seller choose you over another bidder.

Be sure you understand the difference between *preapproval* and *prequalification*. Preapproval is like an estimate provided by the lender, stating the amount for which you're likely to qualify. However, preapproval comes with no guarantees and does not take into consideration your credit score, the property you're financing, and so on.

Although prequalification takes a bit more time and effort, it's important to have a firm commitment for financing before you proceed. If you're prequalified, you can go ahead with your project more effectively and not have to rework your budget, find new (cheaper) contractors, cut corners overall, and worry if your so-called preapproved loan amount turns out to be far more than that for which you can really qualify.

## Whom Do You Choose?

Open the real estate section of your newspaper, and you'll see a collage of photos, promises, percentage rates, and alphabet soup. APR? ARM? Where do you start?

First, understand that you need to shop around for the best deal on your mortgage. Many different types of institutions lend money to people wishing to buy homes. Each has different benefits and may offer significantly different terms on their loans. The following sections present some of your finance institution options.

### Banks

Be sure to check with your local bank when shopping for a mortgage. Whether you have your personal accounts at a savings and loan, a commercial bank, or a mutual savings bank, chances are it offers mortgages. Because some banks either do a great volume in mortgages or, if they're small, have low overhead, there is a possibility you can get the best terms at the same place you deposit your paycheck.

### Credit Unions

According to the National Credit Union Administration (NCUA), a federal credit union is a nonprofit, cooperative financial institution owned and run by its members. These organizations are democratically controlled by their members and offer a safe place to save and borrow at reasonable rates. Members pool their funds to make loans to one another. The volunteer board that runs each credit union is elected by the members. If you have good credit and qualify to participate in a local credit union, this may be a good place to look for a loan. You can find a federally insured credit union at www.ncua.gov.

### Mortgage Brokers

Whereas banks and credit unions represent their own loan options, mortgage brokers represent a wide variety of lenders and loan packages, which they call "products." Depending on the broker's network of contacts and affiliations, he or she may be able to find products that fit various circumstances. Some mortgage companies also have access to private financing or might actually be lenders themselves, so they may have more flexibility to fit your circumstances.

Recently, a crop of online mortgage companies have emerged that offer various packages. You might want to check out these online brokerages for your conventional mortgage, but keep in mind that many of them do not offer construction loan packages.

## Finding the Right Lender

With so many options, how can you be sure you're choosing the right lender? Referrals are usually the best way. Ask friends and neighbors, but also consult your financial planner or accountant. Realtors are also a good resource. Be sure to check out your lender by contacting your state's department of banking as well as your local Better Business Bureau to see whether any complaints have been filed against the loan representative or institution.

Schedule a time to speak with the lender, and ask for the information he or she will need to give you an estimate of the loan for which you'll qualify. Share a copy of your (clean) credit report, and get an estimate in writing of the amount, terms, and interest rate of the loan you qualify for.

Don't give in to pressure tactics, though. If a lender tells you that you need to commit right away, walk away. Although interest rates do change often, you need to protect your interests. High-pressure tactics almost always indicate that a lender doesn't have your best interests in mind.

## Conventional Mortgages vs. Construction Loans

When you open the newspaper, many of the interest rates you see in the bold ads trying to grab your attention won't apply to you. Because you're building your home instead of buying one already constructed, you need a construction loan rather than a conventional mortgage.

Construction loans differ from conventional mortgages because they cover the materials, the labor, and, sometimes, the land you need to build your home. They have a higher interest rate than a conventional mortgage, which is simply a loan you use to purchase a house and property. There's more risk involved in financing the building of a home than financing a structure that already exists—the higher the risk, the higher the interest rate.

When you're shopping for a construction loan, there are generally two types from which to choose: traditional construction loans and construction-to-permanent loans.

### Traditional Construction Loan

This loan pays for the building labor, materials, and expenses in "draws," or installments, that are issued after the work has been

**DON'T TRIP**
on your **SHOESTRINGS**

Don't give out your Social Security number and let banks or mortgage brokerages check your credit multiple times. The credit-scoring system will allow for several inquiries into your mortgage status and not decrease your score, but you may lose a few points if you have many inquiries, especially if you've opened a new credit account or made a financed purchase lately. Instead, take your copy of your credit report—the corrected version, if that was necessary—and offer to share copies of that report with lenders.

done and required inspections completed. These loans usually need to be "bought out" at the end of the construction process. That means you'll likely need to qualify for a conventional mortgage—and have the commitment papers to prove it—before you can qualify for the construction loan.

### Construction-to-Permanent Loans

This loan converts from a construction loan to a conventional mortgage. After you've satisfied the construction loan requirements for building your home, including the necessary inspections, the loan automatically converts to a 30-year, lower-interest-rate mortgage. The rates on the conventional mortgages tend to be a bit higher—you may find yourself paying ¼ to ½ interest point more than if you shopped the loan on your own. However, you avoid the closing costs of the first option, which can total several thousand dollars. Also keep in mind that this type of loan can be more difficult to secure; so if your credit is shaky, if you're self-employed, or if you have a job that's commission-based or based on large bonuses, you may want to opt for the traditional construction loan.

**DON'T TRIP** on your **SHOESTRINGS**

Be sure to adequately estimate the funds you need (see more about setting your budget in Chapter 4). Lenders who deal in construction loans are very familiar with the costs of building a home. If you estimate too conservatively, chances are they'll see that, dub you underqualified to complete the project, and reject your loan. Because you'll be converting to a conventional mortgage after you complete the project, it's better to estimate higher than what you think you need, especially considering that you only pay interest on the amount you draw. When you refinance, you'll only need a loan for what you actually borrowed—not the amount you estimated.

# Construction Loan Considerations

When you're finalizing the terms of your construction loan, you must consider a number of things. Construction loans can be complex, and the rules that govern them are designed to protect the lender's interest. There aren't as many hard-and-fast rules when it comes to construction loans as there are when it comes to conventional mortgages.

The following sections outline some potential pitfalls and issues to be aware of when applying for a construction loan.

### Land Purchase Requirements

Does the loan include the purchase of the land, or does it require that you already own the lot? Sometimes construction lenders want to see that you've already purchased the land, which offers another level of security for the loan. After all, it's likely that there's equity in the land, which would ensure that the lender won't lose money in case the loan goes bad. If you're required to purchase the land first, you can speak with your lender about separate financing packages.

Another option to investigate is the seller financing the land. In this situation, you strike a deal, sign the paperwork, and the seller holds the mortgage to the property and you pay him or her, interest and all, just as you would pay a mortgage company. The seller may take a second mortgage on the property and use your payments to pay back the loan. Or he or she might just bank your monthly payments. In any case, you can often negotiate a better rate with the seller, especially if he or she is motivated to get some cash out of the property.

### Draws or Disbursements

With a construction loan, at certain points during construction, you will receive "draws," or portions of your loan money. But at what times will you receive that money? How much will each draw be? Is that enough to complete the next phase of work, or will you need extra cash or a credit line to get you through? What is required for the money to be released? How soon after meeting those requirements can you expect a check? You need the answers to these questions to be able to work out payment arrangements with your contractors and know at what points you'll have money available to you.

**DON'T TRIP**
on your **SHOESTRINGS**

Be aware that most construction loans are structured so you have to float the initial investment first and then you get reimbursed. For instance, if your loan is for $150,000 and has 5 equal draws or disbursements, you'll likely need to float the money necessary to get to the first draw. In Gwen and Hank's case, our first draw came after the foundation was poured, so we needed to make arrangements with the excavating, masonry, and waterproofing companies to wait for payment until after the foundation was completed. If we weren't able to do that, we would have had to pay these contractors out of our pockets first. Be sure you have enough cash on hand or temporary financing (such as credit cards or lines of credit) to ensure that you can cover the necessary expenses to keep the project moving.

### Permit Requirements

Does the loan require that you have all your permits and approvals in advance of applying for the loan? If so, you will want to work with the municipal building department to see how broad the permit requirements are and see if a preliminary approval is available. Otherwise, you may need to invest more time in making decisions about your structure and choosing a floor plan before you purchase your land. We do not recommend this, however, and suggest that if permit requirements are necessary, you investigate other lending options.

### Contractors

Will the loan allow you to do some of the work yourself, or
do you need to have an approved general contractor (GC)
or subcontractors do the work? If you're required to have a
general contractor oversee the project and you planned on self-
contracting, you may need to adjust your budget to pay for the
additional fees a GC will charge.

### Cost

What are the interest rates, points, and fees on the loan? (See
the following section for more on points.) Are there prepayment
penalties or other costs to the loan that you need to consider?

## Choosing Your Conventional Mortgage

Unless you've opted for the construction-to-permanent option,
after your construction is complete, your construction loan
will need to be "bought out" with a conventional mortgage.
In other words, you'll need to refinance your construction loan
with a conventional mortgage. The process will be very similar
to securing your construction loan, although your rates will
likely be much lower.

Many types of conventional mortgages are available today, so
you'll need to do some shopping to find the right loan for you.
The following sections detail some things you need to keep in
mind as you choose your loan.

### Fixed or Variable Rates

Interest rates change frequently and are affected by a number
of factors. However, you should also pay attention to whether
your loan is offered at a fixed or variable rate (often called an
adjustable rate mortgage, or ARM). A fixed rate stays constant
throughout the life of your loan. A variable rate can fluctuate as

**DON'T TRIP**
on your **SHOESTRINGS**

The longer you keep
your loan, the more
money a lender
makes, so some
lenders assess a
prepayment penalty
if you pay off your
loan before its term
expires. Even if your
state forbids prepay-
ment penalties (many
do), don't assume
you don't have one.
The charter for feder-
ally chartered banks
supersedes their
state mandate, and
they may be per-
mitted to charge pre-
payment penalties.
Before you sign your
loan, be sure your
lender doesn't assess
a prepayment
penalty.

market interest rates fluctuate, but often will be "capped" to stay within certain parameters during one year.

You might also find hybrid loans, which may remain at a fixed rate for a period of years and then convert to annual ARMs.

Variable rates can save you money if interest rates go down or can cost you more money if interest rates go up, so bear the market conditions in mind when you choose your loan. If you can lock in your loan at a low fixed rate, that's usually the best way to go.

### Balloon-Payment Mortgages

The payments on these types of mortgages are calculated as if they're traditional 15- or 30-year loans, but the balance of the loan is due in one large sum (the balloon payment) after 5, 7, or 10 years.

Be sure you evaluate the loan offering based on your individual situation, payment abilities, and payment habits. Overall, look for the lowest rate for the longest possible time you can get, and pay off your principal balance as efficiently as possible.

### Interest-Only Loans

These loans seem like a great deal, with their low interest rates and low monthly payments, but be careful of them unless you are extremely financially disciplined. When you pay this loan each month, you are paying only the interest on the principal balance. That means you're not actually paying back any of the money you borrowed—only the interest on that money. You could pay your loan every month for 30 years and still not own your home at the end of that time. If you do opt for an interest-only loan, be sure you pay extra every month to ensure you will be paying off your principal balance and building equity in your home. If your budget is continually stretched tight, however, and you think you might be tempted to only make the interest payments, you may be better off opting for a higher monthly payment that includes principal reduction.

On the other hand, if you have a seasonal job that allows you to pay more on your mortgage during some months, while other months are lean, this could be a good option. In addition, if you're in a field where your income is likely to greatly increase over time, such as an attorney or a doctor just starting out, these loans could offer a way to get more house for your money. Again, for it to be worthwhile, it's critical that you exercise discipline and pay down your principal.

# Government Loan Programs

Folks who have trouble qualifying for traditional financing may find other options through government loan programs. In these cases, the government agency itself does not directly loan money. Rather, the agency insures particular loans against default, offering lenders an additional measure of security and making it possible for those who might not qualify for a loan to do so or for an applicant to receive better terms. Two options that you may want to investigate are the Federal Housing Administration (FHA) and the Veteran's Administration (VA).

### Federal Housing Administration
Founded in 1934, the FHA insures loans made by FHA-approved lenders throughout the United States and its territories. This can be an option for people who would otherwise have trouble getting a loan, but be aware that their inspection requirements are a bit stiffer than many conventional loans. Ask your lender representative if he or she offers FHA-insured loans.

### Veterans Administration
Similar to FHA-insured loans, the VA insures loans for members of the military. If you are a veteran or currently on active duty, this may help you more easily qualify for a loan or get a better rate.

## MONEY IN YOUR POCKET

Don't spend money for expensive kits or programs that promise to help you pay off your mortgage early by making bi-weekly payments. It's true that paying your mortgage bi-weekly—paying half the amount every 2 weeks instead of the traditional monthly payment—can save you quite a bit of money. In addition to paying one extra payment a year (paying every other week totals 13 full payments), you're also reducing the principal balance—the core amount lent to you— more quickly and can reduce the term of your loan by 7 years.

That extra payment a year makes a big difference. For example, someone who borrows $100,000 at 6 percent interest for 30 years pays approximately $600 a month in principal and interest. Taxes and insurance make the payment about $1,000. By paying $500 every 2 weeks, which equals one extra $1,000 payment per year, you could pay off your loan in a little more than 22 years, saving nearly $34,000 interest.

**Savings for You: $34,000**          **Running Total: $34,000**

## Common Closing Costs: What's Negotiable?

Beyond your down payment and monthly principal and interest payments, your loan will have a number of costs associated with it. You should request—and your lender should give you— a good-faith estimate of the fees and expenses associated with the loan. In fact, many states require this by law. Few lenders will charge all the fees listed here, but you should still receive a good-faith estimate that documents what (and how much) will be charged to you. Fees and costs associated with your loan may include the following:

- **Loan-origination fee.** This fee can also be called a "point" or "points" and represents a percentage of the loan charged to you as a cost of producing the loan. One point is equal to 1 percent of the loan value. This fee is usually negotiable, and some lenders don't charge points at all. Points can also be used to lower the interest rate on the loan in the long term.

- **Appraisal fee.** This is a fee for a licensed real estate appraiser to inspect your property and compare it with other homes on the market to determine its value. Your appraisal is different from your tax assessment, and the value may be greater or lower than your assessment. This is usually not negotiable.

- **Credit report.** This fee, paid to a credit reporting bureau to access your credit report, is usually not negotiable.

- **Lender's inspection fee.** This fee is common in construction loans and covers the cost of a lender's representative to visit the property and inspect the work. This may be negotiable, depending on the loan and the lender.

- **Mortgage broker fee.** A fee paid to the mortgage broker above and beyond the commission he makes for brokering your loan. This is often negotiable.

- **Underwriting fee.** A fee the lender charges to cover the cost of having a professional underwriter review your loan application. This is often negotiable.

- **Wire transfer fee.** This covers the cost of wiring money to or from your account to complete the transaction.

- **Application fee.** This fee is from the lender and covers the cost of reviewing your application. It is often negotiable.

- **Commitment fee.** This is another fee charged by the lender for securing a loan commitment. It's often negotiable.

- **Flood certification fee.** This fee is for a lender's representative to verify whether your property is located in a flood zone. If the property is in a flood zone, it may affect the financing, or you may be required to secure flood insurance. It's usually not negotiable.

- **Express mail or courier.** This is a fee for any delivery of documents to and from the attorney, seller, or other entities. It may be negotiable.

- **Closing or escrow fee.** This fee, charged for closing the loan, is often negotiable. It's usually not applicable if an attorney is used to close the loan, because that cost is usually part of his or her fee.

- **Document preparation fee.** This fee covers the time and expense of preparing and copying documents related to the transaction. It may be charged by the lender or the broker.

- **Notary fee.** This is the fee for having a notary public certify your closing documents, which the lender usually requires. Lenders often have a notary public on-site, so this fee is often negotiable.

- **Attorney fees.** This is the fee for your attorney's services, if you used an attorney. When you hire an attorney, always negotiate the fee for service up front and require that any changes to that fee be approved by you in writing first. That way, you won't have surprises at the closing table.

- **Title insurance.** This insurance policy ensures that the title to your home is "clean," meaning it has no liens against it and that the seller is entitled to sell you the property free and clear. This is usually not negotiable, although if the seller has a title policy from a previous purchase, you might get a "reissue" rate, which is usually less expensive. In the case of a traditional construction loan, you should get the refinance rate on the conventional mortgage, which will save you even more money.

- **Recording fees.** This fee covers the costs of filing your sale and its details with the appropriate municipal and county governments. It's usually not negotiable, but in some cases, it may be included in other fees.

- **City/county tax stamps or state tax stamps.** This fee simply means any appropriate city, county, or state taxes associated with the sale. Although not all areas have such taxes, when they are required, they are usually not negotiable.

- **Pest inspection.** The lender may require that the home be certified to be free of termites or other pests. This is usually not negotiable, but you may be able to have it done on your own and save some money.

- **Survey.** This fee is the cost of having a professional surveyor visit the property and certify the land boundaries. It's usually not negotiable, but be sure to request and save a copy of your survey. When you secure a conventional mortgage soon after the construction is complete, you might not have to pay to have another survey done if you can produce one that was recently completed. At the very least, you can ask the survey company to recertify a recent inspection for a reduced fee.

- **VA funding fee.** This fee is charged only to loans issued by the Veterans Administration.

In addition, you may be required to be pay some items in advance, such as the following:

- Interest for 30 days or from the day you close until the end of the month.

- Mortgage insurance premium (also known as private mortgage insurance, or PMI)

- Hazard insurance premium (also known as homeowner's insurance)

Several reserves will often be deposited with your lender. This money is held "in escrow," which means that the bank holds the money in an account until these expenses are due. This money is used to cover certain expenses associated with your loan, including the following:

- Hazard insurance premium*
- Mortgage premium reserves*
- School tax
- Taxes and assessment reserves
- Flood insurance reserves*

*(\*Usually 1 month's premium, but check with your lender to be sure.)*

# GO FIGURE

How much can you save on your loan's closing costs? Use this worksheet to tally the fees and charges that will be assessed to your loan. Be sure to put those big, fat zeros in the categories that you're able to get waived!

Loan origination fee _____
Appraisal fee _____
Credit report fee _____
Lender's inspection fee _____
Mortgage broker fee _____
Underwriting fee _____
Wire transfer fee _____
Application fee _____
Commitment fee _____
Flood certification fee _____
Express mail or courier charge _____
Closing or escrow fee _____
Document preparation fee _____
Notary fee _____
Attorney fees _____
Title insurance _____
Recording fees _____

City/county tax stamps or state tax stamps  _____  _____
Pest inspection  _____
Survey  _____  _____
VA funding fee  _____
Other:

_____

_____

You might feel as if you're signing your life away at the closing table. Be prepared to sign between 20 and 30 documents, depending on your state's requirements. Sometimes attorneys and lenders will rush through this process. Don't let them. Be sure you understand every document you're signing. Don't be afraid to ask questions—remember, you're the customer. You shouldn't sign any document unless you're fully aware of what it states.

## MONEY IN YOUR POCKET

You can almost always get the application or origination fee waived on your loan if you push hard enough. You can also ask for interest rates to be reduced. With the waiver of a $250 application fee and reducing the points you need to pay up front by ½ point on a $200,000 loan, you've put some cash back in your pocket.

| **Savings for You: $1,250** | **Running Total: $35,250** |
|---|---|

# Secrets Lenders Don't Share

As you shop around for your loan, it's important to keep one mantra in mind: everything is negotiable. Well, *almost* everything is negotiable. What many people don't realize is that they can negotiate with lenders, asking for a lower rate or for an application fee to be removed. The following paragraphs list some of things lenders don't generally tell you:

Although negotiating can save you money, keep in mind that the final deal has to be a win-win situation. If you're demanding a rate significantly below prevailing rates or refusing to pay necessary fees, such as those for appraisals, attorney services, and flood certifications, the lender is not likely to be motivated to help you. Do your homework about what you can reasonably request, given your situation, by looking at rates, speaking with others who have secured similar financing, and speaking with various lenders about what they offer. That way, you can ask for a deal that's fair and equitable for both you and the lender.

- **You don't need a 20 percent down payment.** In most cases, programs are available that require a significantly lower down payment—sometimes as low as 3 to 5 percent. So although it's always a good idea to save your pennies, keeping them more liquid by putting less money down can be a smart strategy. You'll have a cushion of cash put aside in case expenses run a bit high or you need extra financing between disbursements.

- **Lenders want your business.** It's easy to put lenders up on a pedestal, but remember, *you* are the customer. They make money by selling loans to people like you. Don't be afraid to ask for the best deal.

- **Interest rates are often negotiable.** Just as you would ask for a better deal on a car than what's on the sticker, you might be able to reduce your interest rate just by asking. Some lenders price their own packages, much like a car dealer prices its vehicles. There's usually room to negotiate, especially if you meet the qualification criteria.

- **Some fees can be eliminated.** Ask questions about every fee that's reflected in your estimate. Often such costs as the application fee can be eliminated. Find out if you can save money by having your own flood or termite certification done, instead of going through the lender's provider.

- **Bigger isn't always better.** Sometimes your local bank can give you a better deal than a large mortgage broker representing dozens of loan products. Be sure to investigate all your options.

## The Paper Chase

It might seem like you need a forest's worth of paper to apply for a mortgage. Indeed, the documentation can be steep, and different lenders require different proof of your ability to repay their loans. Tracking down and organizing your paperwork in advance can make the application process much smoother. The following sections discuss the documentation you're likely to need.

### You and Your Income

Lenders consider your annual income when measuring your ability to repay a loan. They also look for stability—in your job, income, and residence. Moving residences or changing jobs several times within a couple of years can be a red flag, so be sure to explain any extenuating circumstances to your lender. Be able to present the following information and paperwork:

- Your name, address, phone number, and other contact information. If you've moved in the past 2 years, provide your previous address(es).

- Your marital status and information about any co-borrowers.

- Your Social Security number, as well as that of any co-borrowers.

- Proof of citizenship, if you are a naturalized American citizen.

- Employment information for the past 2 years, including the dates you were employed and gross monthly income.

- W2 forms for the past 2 to 3 years and one or more of your most recent pay stubs.

- Proof of any other income sources, such as alimony, pensions, dividends, or Social Security.

- If you're self-employed, you may need to provide your business and personal tax returns for the past 2 to 3 years, including all schedules. Some lenders may require a year-to-date profit-and-loss statement certified by your accountant.

## Your Assets

Lenders want to see that you have accumulated some assets. You don't have to be rich—they just want to verify that you're not living paycheck to paycheck, which may inhibit your ability to repay the loan. They also want to see that any down payment money has been in your account for several months. If the sum has been deposited recently, you need a letter explaining where the money came from. If it's a gift from a relative, you need a letter stating that. Lenders don't want mortgage applicants borrowing money from friends or relatives to beef up their bank statements. Have the following information available:

- Three months of bank statements for your checking and savings accounts
- Statements of any other financial holdings, such as stocks, bonds, mutual funds, IRAs, and others

## Your Debt

Lenders want to see how much debt you're carrying in relation to your income to ensure that you have adequate resources to repay the loan. They also look for the amount of credit you're carrying in relation to the amount of credit you have to ensure that you're able to responsibly manage your debt. Have these documents ready:

- Information about your current mortgage (if applicable), including the lender's name and address, and the most recent statement.
- Proof of payment for 12 months (such as cancelled checks or statements showing payments credited) of your mortgage or rent.
- Current creditors, including credit cards, car loans, student loans, etc.

- If you rent your residence, the lender may want to see 12 months of cancelled checks or rent receipts. A letter from your landlord stating a history of prompt payment may suffice.

## Past Credit Problems

It's a common assumption that people who have declared bankruptcy in the past can't get mortgages. Although it's true that your interest rate and, perhaps, your down payment will likely be higher, it's not impossible to get a loan after a bankruptcy. If you've had credit problems, the lender wants to see that there's a reasonable explanation or that the credit problems have been resolved. Be able to present the following information:

- Copies of any petitions for bankruptcy, the schedule of bankruptcy, as well as the discharge of bankruptcy.

- Any judgments against the loan applicant will require a copy of the release or satisfaction of the decree for the lender.

- Copies of any divorce decrees, proof of any name changes, and transcripts if you were a full-time student within the past 2 years.

- For late payments or other problems, a letter of explanation will often help your case. If you missed a credit card payment or were late on your car loan because of an illness or death in the family, write a letter explaining that. If the rest of your credit is pretty clean, it will help the lender understand that the issue was due to extenuating circumstances.

# MONEY IN YOUR POCKET

If your loan requires you to buy private mortgage insurance (PMI)—usually for homes with less than 20 percent equity—you might be able to avoid buying the insurance using what's commonly called 80-10-10 financing.

If the lender has to foreclose on a property, PMI protects the lender. If the borrower can't pay back the loan, the lender needs to be sure there's enough equity in the property to cover the costs associated with reselling the property to recoup the debt. Those costs include such things as attorney fees, taxes, etc. To be sure the lender is covered, they'll look for insurance for any loan greater than 80 percent loan-to-value (the amount of the loan versus the value of the property). However, different loans require different levels of insurance.

Some lenders will allow you to take a mortgage for 80 percent of the financing you need and then take a home-equity loan on the property for 10 to 20 percent of the financing, depending on your down payment. That eliminates the need for PMI and can put approximately $30 to $100 per month in your pocket, depending on the size of the loan.

Keep in mind that the financing rate on the home-equity loan may be slightly higher than the mortgage amount, and you will need to pay two loans. But you'll be paying more on the equity amount of your loan instead of wasting money on insurance, so you'll be building your asset base more quickly.

If this is not an option for you, and you still need to pay PMI, watch the market and your principal payments closely. If your home appreciates, or if you're able to pay a bit more on your mortgage, you may be able to get out of the PMI requirement more quickly when your loan-to-value percentage reaches 80 percent. You usually have to request that it be removed.

**Savings for You: $7,200**             **Running Total: $42,500**
*(equal to $20 per month over*
*the life of a 30-year loan)*

## At the Closing Table

You're not likely to ever have to sign as many papers in one sitting as you are at a mortgage closing. The closing usually takes place in the office of your or the seller's attorney, and the seller may or may not be present. Different states and different lenders have different paperwork requirements, but they generally fall into a few broad categories:

- Confirmation of your loan and its terms
- Reconfirmation that the information you supplied to get the loan is accurate
- Details about the property
- Information required by law to be given to you

We recommend reading each document, or at least skimming each document and ensuring that you know what you're signing. Double-check all figures and calculations—we've been at closing tables where fees we had negotiated to be waived were on the closing statements, meaning we'd have to pay them. Mistakes do happen, so it's critical that you be sure the rate and terms are what you negotiated. If they're not, speak up and insist that the errors be changed. It's possible this can be done on the spot, or you might need to schedule a new closing date to allow time for the corrections to be made on the documents by the closing attorney or the lender. This may be frustrating, but it will be worth it in the long run when you put that extra cash in your pocket.

If you feel as if you're being rushed, remind the people who are rushing you that you are the customer. You're the one buying the property. You're the one being serviced by the attorney. Don't be afraid to ask questions if you don't understand something. These are legally binding documents, and, when they're signed, you're responsible for the terms and commitments made, regardless of whether you understood them at the time or not. This is likely to be the biggest investment you'll make in your lifetime. It pays to take the extra time to be sure everything is correct and you're knowledgeable about the documents you are signing.

## Dollar-Saving Do's and Don'ts

- Prequalification gives you more leverage to negotiate when you're shopping for property.

- Shop around for the best lenders, and remember that most fees and interest rates are negotiable. Asking for a better deal will usually save you money.

- Even people with negative information on their credit reports can get loans, although they may be offered at a higher interest rate.

- Don't let multiple lenders run your credit report, as that could decrease your score. Obtain a free copy of your report to share with lenders and then let them make an offer based on that.

- Get your documents in order well before you apply for your loan, and keep extra copies of everything. This will make the process of obtaining your loan much less stressful—and could save you time and money!

# 3
# CHOOSING YOUR LOCATION

*Where* you build your dream home is just as important a decision as *what* you build. Finding the right property is critical to ensuring that your dream home will be one that suits your lifestyle and your needs for many years to come. So don't limit yourself with preconceived notions of where you think you'd like to live (or not live). Look for a property in a location that will suit your needs and lifestyle, but remain flexible so you find the piece of property that suits you best.

Before you go out lot shopping, take some time to think about what's important to you. Do you particularly like certain towns or areas? Do you need to consider the quality of schools? Do you need to be close to major highways or mass transit for commuting to your job? Do you want to live in an area where many services are available, or does a more rural lifestyle suit you? Don't just choose a lot because of its size or price. Do your homework when evaluating a piece of land. This chapter shows you how.

# GO FIGURE

In Chapter 1, you briefly noted in what area you'd like your home to be. Here, you get more specific about what you need in a piece of land.

| | Very Important | Somewhat Important | Not Important |
|---|---|---|---|
| *Neighborhood/Area:* | | | |
| Urban | ☐ | ☐ | ☐ |
| Suburban | ☐ | ☐ | ☐ |
| Rural | ☐ | ☐ | ☐ |
| Quiet | ☐ | ☐ | ☐ |
| Active | ☐ | ☐ | ☐ |
| Parks, recreation nearby | ☐ | ☐ | ☐ |
| Quality and proximity of local schools | ☐ | ☐ | ☐ |
| Style/condition of other houses nearby | ☐ | ☐ | ☐ |
| | | | |
| *Lot:* | | | |
| Size | ☐ | ☐ | ☐ |
| Condition | ☐ | ☐ | ☐ |
| Buildable area on lot (are there wetlands or other restrictions?) | ☐ | ☐ | ☐ |
| Soil quality | ☐ | ☐ | ☐ |
| Drainage | ☐ | ☐ | ☐ |
| Tree coverage | ☐ | ☐ | ☐ |
| Location | ☐ | ☐ | ☐ |
| Proximity to other homes | ☐ | ☐ | ☐ |
| | | | |
| *Convenience Factors:* | | | |
| Proximity to mass transit | ☐ | ☐ | ☐ |
| Proximity to services (hospital, doctor, dry cleaner, etc.) | ☐ | ☐ | ☐ |
| Proximity to shopping/amenities (supermarket, etc.) | ☐ | ☐ | ☐ |
| Availability of public water | ☐ | ☐ | ☐ |
| Availability of public sewer | ☐ | ☐ | ☐ |
| Availability of natural gas | ☐ | ☐ | ☐ |
| Other features: _____ | ☐ | ☐ | ☐ |
| _____ | ☐ | ☐ | ☐ |
| _____ | ☐ | ☐ | ☐ |

# Which Land Is Your Land?

As you get a clearer picture of the lot that you have in mind, you should be aware that some often overlooked aspects of land can cause unhappy surprises if they're discovered after the deal is sealed. Consider the questions in the following sections while you lot shop.

### How Will Your Dream Home Fit on This Property?

Be sure the size of your lot will accommodate the size of house you want. It's probably not a good idea to build a 4,000-square-foot home on a tiny lot, even if that lot is in a great neighborhood.

### What Is the Area Like at Different Times?

Sure, the neighborhood might be peaceful if you visit during the week when all the local kids are in school, but what does the lot sound like at different times of the day and different days of the week?

Residents of the town in which Gwen and Hank live had an unwelcome surprise when they moved into their development of new estate homes and found out that a nearby speedway hosted a cadre of incredibly loud race cars three evenings a week. Find out if your prospective lot has any noisy or otherwise unpleasant neighbors, including race tracks, airports, garbage dumps, train stations, power plants, and the like, which may affect your quality of life—and the value of your property.

### Is Traffic a Factor?

Much like you want to check for the noise factor at different times, if traffic is important to you, visit the property during rush hours and on weekends to check the volume of traffic at those times.

### Are These the People You Want as Your Neighbors?

If you like to look out your window and see neatly kept lawns, don't buy a property across the street from a junkyard. Sometimes property buyers get so excited by one aspect of a piece of land they end up ignoring factors that may really begin to bother them over the long term. Pay attention to whether you'd really like to live in this area.

It's also a good idea to go and knock on the doors of some of the people in the neighborhood. Ask their opinions about the area and whether they like living there. You will be surprised at how willing some people are to share honest opinions if you simply ask. This will also give you a good feel for how friendly and helpful the people in the area are.

### What Will Your Home Look Like on the Property?

If you long for a Spanish villa but choose a seaside lot in Cape Cod, your home is likely to look out of place, which could affect its resale potential. Be sure your dwelling will fit the style of the neighborhood.

### What Size Yard Do You Want?

Visions of rolling acres can be nice, but a big yard is a whole lot of work. Do you really have the time or inclination to landscape and mow an acre or two, or is a smaller lot more manageable? Consider the extra costs involved in owning a big lot, including landscaping and watering the lot. Keeping a couple acres of property green can boost your water bill significantly.

### Can You Build Your Dream House on the Property?

Some pieces of land come with deed or zoning restrictions that may affect what you are permitted to build. Be sure the land doesn't have any hidden usage surprises. In addition, some areas have so-called paper roads, or road extensions that could be made in the future. Be sure you don't have paper roads near your property, or one day you could find that your out-of-the-way home is a little too close to a new street.

Also because the resale value of your property should always be a consideration, it's a good idea to follow the old adage of buying the best piece of land in the best possible area you can afford. Even if you believe you'll never leave the home you build, it creates more value in your property for the future.

## Finding Your Lot

After you've covered the big-picture considerations of your land, you can evaluate the properties that fit your parameters.

### Home Base

When you look at your lot, determine where the structure would need to go. Gwen and Hank had an unpleasant surprise when they thought they'd found a great lot but learned that the coverage requirements—the percentage of the property that could be covered by buildings, driveways, and other structures—severely limited which house plans they could choose. They moved on to find another property.

Also check the direction your house will face. Northern exposure means you'll get less natural light through your windows, and more natural light can help you save on your electricity and heating bills. If that's important to you, be sure you choose a lot on which your home will face east or south.

### Soil Quality

You might not think too much about the quality or location of the soil on your lot, but it is an important consideration for a number of reasons:

- **Adequate drainage.** If your lot is in a low-lying area or soil quality is claylike, rainwater may not drain well,

Be sure your lot isn't near unseen hazards. Most states have Superfund sites, or sites the government has determined are contaminated with toxic materials. Check to see if such sites exist near where you want to buy by visiting www.epa.gov/superfund/sites. Call the Environmental Protection Agency (EPA) or your state's Department of Environmental Protection or similar office to find out if there are other environmental issues in the area.

which could cause flooding problems both on your property and in your basement, if you decide to include one. Poor drainage can also be a problem for homes that require septic systems. Conversely, sandy soil generally drains well and may mean that your property is less likely to flood. However, soil that is too sandy may be unstable.

- **Excavating.** Hard, rocky soil can be more difficult to clear and dig, adding to your excavating costs. In the most extreme cases, when a property has a base of rock, the only way to dig a foundation is to use dynamite, which is clearly not an option if your neighbor's home is 50 feet away. Also be aware that wooded lots will need to be cleared, and you'll likely incur extra costs for digging out roots and stumps.

- **Slopes or uneven lots.** If your lot is sloped or has a hilly landscape, you could end up paying more for excavating, because it may need to be leveled before you can build. Many municipalities also have restrictions about how the property has to be sloped, so water runoff from one property doesn't flood another. If these slopes don't occur naturally, you may be liable to create them—again adding more to your cost.

- **Land use.** If you are an avid gardener or have other plans for land use that rely on the richness of the soil, pay particular attention to your property's soil makeup.

- **Top soil.** If your lot doesn't have enough topsoil or the topsoil isn't good-enough quality, you may need to have more topsoil trucked in—at an additional cost.

### Water Table
Your lot's water table is the level at which water exists under the surface of your topsoil. The higher the water table, the more

challenges you are likely to have with building on the property. Some areas have such high water tables that basements are not possible, because they would flood continually. Significantly high water tables may require that water be pumped away from the building project—again adding to your cost.

## Utility Services Available

Some lots have all the luck—access to public sewer and water as well as natural gas hook-ups and other services. Check to see which are available on the lot you are thinking of building on. If you don't have access to public water and/or sewer, you may have to invest in well and septic systems. In some cases, local natural gas providers charge to run access lines to lots that don't have them. If you have definite preferences about the utilities and services you want to use, be sure the lot you choose will have them.

## Planned Community Requirements and Homeowners Associations

A number of communities are governed by local homeowners associations, which can place restrictions on property use. If you're purchasing a piece of property in a planned community or an area that has a homeowners association, be sure to familiarize yourself with building requirements, as well as any lifestyle restrictions. You don't want to build your dream home and then find out that you're going to get fined every time your child leaves a bicycle in the driveway or if your grass grows a bit too high.

## Historic Districts

If the lot you're buying is in a historic district, or if you're planning on purchasing a lot with a home on it and demolishing the home to make way for your new structure, be sure there are no historical restrictions on your doing so. Historic districts may have strict parameters in place that specify what type of home you can build—sometimes even dictating the exterior house color.

### Easements
An easement gives a third party some right to or restriction on your land for a specific purpose. For instance, if you have a corner lot, it may have a sight easement that requires that you keep the first 10 feet from the street clear of all obstructions, including fences and bushes. A nearby lot may have an easement on your property that allows it to use part of it for access, such as having part of a driveway on your property. Any easements or other usage requirements should be spelled out in your deed.

## A Lot for Less
Shopping for land can be a great adventure—as long as you're armed with the specifics. You can find lots for sale from a number of sources, and some will provide you better opportunities and prices.

### Newspapers
Your local newspaper probably lists real estate sales in its classified section every day. Keep an eye on local paper advertisements, especially on Fridays and Sundays, which are usually heavy real estate advertising days.

### Private Sales
With private or "for sale by owner" (FSBO) sales, an individual or entity sells a piece of land without the use of a Realtor. This may be a piece of land the owner wants to get rid of, or it could be a subdivision—the parceling off of a piece of land from an existing lot. In the latter case, you need to be sure the subdivision is

approved for building. The seller of the property should be able to give you documentation of the approvals, or you can check with your town's building office.

Because a real estate agent is usually not involved in these transactions, the seller saves the cost of commissions, which could be as much as 6 percent of the sale price. You may have more room to negotiate the property price if you're dealing directly with the seller.

### Land Auctions

Check the "Public Notices" section of your newspaper to find land auctions in your area. This could be publicly owned land, land that has been on the market for a lengthy time, or land that has been seized from the previous owner for a variety of reasons. It's always a good idea to have a firm knowledge of the land's value and also your budget. That way, you won't be tempted to overbid if there's a great demand for the property. However, these land sales don't often draw high attendance, so it's often possible to walk away with a great deal.

### Builder Lots

Sometimes builders offer good prices on odd-size lots or lots in existing developments that didn't sell for one reason or another. This could mean a bargain for you and quick cash for the builder. Before you buy, check to be sure there are no issues related to building on the property.

## The Realtor Deal

Another way to find suitable lots is to use a real estate agent, or Realtor. The two terms are often used interchangeably, but a real estate agent refers to a person who is licensed or otherwise

**DON'T TRIP** on your **SHOESTRINGS**

Don't assume that the best way to save on a lot is to down-size. Although less land may be less expensive, if your lot is too small, construction crews may have trouble navigating their equipment on it, and you may end up paying extra for such things as hand-trucking of materials and special pump trucks for pouring your foundation. Small lots can be beautiful, but be sure yours is buildable, too.

authorized to sell property; a Realtor is a real estate agent, broker, or associate who holds an active membership in a local real estate board that is affiliated with the National Association of Realtors.

The benefits of using a real estate agent include the agent's ...

- Ability to research properties for you.
- Access to databases of property listings.
- Familiarity with properties in his or her primary market area.
- Knowledge of the area, including everything from schools to taxes to soil quality and everything in between.
- Ability to present you with a variety of possible properties.

Real estate agents serve as guides through the property-finding process and may also have good recommendations for mortgage sources. They can save you some time and help you avoid some headaches.

## MONEY IN YOUR POCKET

Few people realize that you can negotiate realty commissions. With a standard 6 percent commission, half of that commission goes to the buyer's brokerage and half goes to the seller's brokerage, with half of each of those trickling down to the agent. If you're listing your current home with an agent and using that agent to find a lot, that agent is making a whopping 3 percent of the total transaction price from you. If you sell your home for $150,000 and buy a lot for $100,000, the agent is making $4,500 for simply listing the property and another $3,000 when you buy your lot. If the agent is representing both properties, ask him or her to reduce the brokerage's overall commission by 1 percent and you'll save $1,500 on the sale of your home.

| Savings for You: $1,500 | Running Total: $43,950 |

However, real estate agents work solely on commission, so their services do cost you money. Because the combined commissions of agents representing the buyer and the seller can be as much as 6 percent, you have less bargaining power on the price of your property because the seller is earning less on the sale. And although agents are required to disclose whether they are representing the buyer or the seller or both in the transaction, the fact is, it's still in their best interest to be sure the sale closes.

Real estate agents also may not have access to or be inclined to lead you to private sales, because these are not generally listed in databases and often offer no commission to the agent. Be sure to keep up on the listings in your newspaper, even if you are using an agent.

## MONEY IN YOUR POCKET

Services such as eRealty.com, ZipRealty.com, flatfeelisting. com, LendingTree.com, Foxtons (www.foxtons.com), Catalist Homes (www.catalisthomes.com), and others offer discounts on the commissions they pay and receive, passing the savings along to you. Some of the larger discount realty companies have local real estate representatives who can help you with your land selection process. ZipRealty, for instance, rebates you 20 percent of its commission. So buying a $100,000 property would earn you back up to $300, if ZipRealty's commission was 3 percent.

**Savings for You: $300**          **Running Total: $44,250**

# Uncovering Hidden Costs

If you're not careful in your research, you could get zapped with hidden fees and expenses that add to your budget. Watch out for the following common costs.

### Special Assessments

Some towns and counties levy a fee or mandatory "contribution" for various reasons. In New Jersey, for example, a contribution of 2.5 percent of the assessed value of a land purchase is due as a "Mount Laurel Contribution," which subsidizes low-income housing in the town. You can usually find out if your land is subject to such costs from your town's land use office.

### Flood Zones

Many lenders require certification that the property is not located in a flood zone. If you do purchase land in a flood zone, your lender is likely to require that you secure a special flood insurance policy, which can cost hundreds of dollars per year. Be aware of the flood zone issue as you look at properties located near bodies of water or in low-lying areas.

### Road Opening Fees

If your construction will require the extension or opening of a new road to access your property, you may be liable for some or all of that cost.

### Variances

Some pieces of land are restricted by either the zone in which they're located or specific restrictions written into the deed of the property. If your lot is one acre and the area in which it's located requires that houses be built on 2-acre-minimum-size lots, for instance, you'll likely need a variance. Similarly, if your property is restricted by its deed in terms of what can be built on the property and your plan isn't allowed, you may need a variance.

In simple terms, a variance is permission from the town to use land in a manner that is not consistent with its zoning or building requirements. Each municipality has its own methods of obtaining a variance, and you should become familiar with them if you intend to apply for one. Visit the building or land use offices of your town, and ask them if there is a precedent for what you want to do. That can be a good indication that your variance will be approved. For instance, if you know that other homeowners have been allowed to install fences beyond their setback requirements, you'll likely have a better chance than if this request has been repeatedly denied.

You may also need an attorney to navigate the process. If that's the case, choose one who is versed in the zoning regulations and processes of the town in which you're building. Ask building officers in your municipality for recommendations, or check zoning board or land use meeting minutes for names of attorneys handling variances.

Variances can be expensive, and there are no guarantees that even a request that has a precedent will be approved. So avoid them if possible, or shift the onus of getting one onto the seller, if possible.

## Height Restrictions

Know the maximum height allowance for new dwellings in your town. And be sure to find out whether that measurement is from the base of the house or from street level. You don't want to incur additional costs to have your home plan modified if it's slightly taller than what's allowed.

If your lot has restrictions out of sync with what you want to build, you may still be able to make it happen. By making an application to the zoning board, which usually requires that you appear in front of the board and that you hire an attorney, you may be able to obtain a variance, or a waiver of the zoning requirements that allows you to build your desired structure. Applications for variances, plus the legal fees assessed, can run into the thousands of dollars, and there's no guarantee your variance will be approved. You can find out if you're likely to need a variance when you visit your local municipal offices.

## MONEY IN YOUR POCKET

If your desired lot requires a variance for building, you might be able to make the owner responsible for obtaining the variance as a condition of the sale. Shifting the responsibility off you eliminates your financial risk. Legal fees for a variance can run from a few thousand to tens of thousands of dollars, based on the complexity of what's required. If you need a variance for putting a home that would exceed setback requirements on the lot, for instance, your application fee in our town would be $300 and attorney fees would likely be in the neighborhood of $2,500. By shifting that responsibility to the seller, that $2,800 comes out of the seller's profits, and not your pocket.

**Savings for You: $2,800**          **Running Total: $47,050**

## Visiting Town Hall

You can find a great deal of information about a property at its town's municipal offices. Taking a trip to town hall can give you valuable tips, warnings, and negotiating power, so be sure to schedule some time to head to the following offices:

- **Zoning office.** The zoning office tells you the building restrictions for the zone in which your property is located. The area may be zoned for residential or commercial use, or may have limits on the minimum acreage on which a new dwelling can be built.
- **Building office.** Here you can learn about the permits you need before you can build, any setback requirements, and

any additional assessments that may be levied. If the property has special requirements, such as the provision of road access or other necessities, you can learn about that here as well.

- **Land use office.** Sometimes combined with the building office, this is where you can find out about the process of building, including the inspection and approval processes.

- **Tax assessor's office.** This is where you can get an idea of the tax rates in the area, which should be a major consideration before you buy a piece of property. You can also find out about recent sales of comparable properties in the area.

- **Fire inspector's office.** Find out what fire detection equipment is necessary in your home. You can also find out any fire-related restrictions on building your home, such as increased window size to allow firefighter access to second-floor rooms.

The municipal clerk can lead you to the right department to find out more helpful information, such as the following:

- Location of flood zones within the municipality
- Demographic information about the area
- Crime statistics for the area, including which areas have the highest crime level
- Information about schools
- Information on ordinances and other legal issues within the town
- The number of new home permits issued within a given period of time
- Water-quality reports

# Doing Your Due Diligence

After you've decided on a lot, you should order several tests to ensure that you're not going to get hit with any nasty surprises later.

### Survey

A survey marks the boundaries of your lot and is completed by a professional surveyor. Your lender may require a survey as part of the financing requirements, and the cost can range from a few hundred dollars to more than a thousand dollars, depending on the complexity and size of the property. However, this ensures that you know exactly where your land ends. Don't assume that just because the seller says the land boundaries are in a specific location that those boundaries are accurate.

## MONEY IN YOUR POCKET

Before you schedule a new survey, which could cost as much as $500, see if the lot's owner has an existing survey. If so, you might be able to simply get a recertification of the existing survey. You may save as much as half if you opt for recertification.

**Savings for You: $250**          **Running Total: $47,300**

### Water Test

This is less of an issue when you have city water; if you're going to need a well, however, you'll need to know that the water beneath your property is free of contaminants. Different areas have different water levels, so you'll also need to know how far down you need to drill for your well to get an adequate supply and quality of water.

### Soil

A soil test is essential to determine the quality and composition of your soil. If you're going to need to install a septic system, most municipalities will require that the lot have a percolation—or perc—test done to determine the rate at which the soil absorbs liquid. The result will impact the feasibility and type of septic system required.

If the owner hasn't provided these tests as part of a subdivision process, your agreement to buy the lot should be contingent upon these tests yielding successful results.

## GO FIGURE

Use this worksheet to determine the obvious—and not-so-obvious—costs of your property.

| | |
|---|---|
| Purchase price of land | _____ |
| Closing costs | _____ |
| Survey | _____ |
| Soil testing | _____ |
| Water testing | _____ |
| Grading | _____ |
| Clearing | _____ |
| Rough driveway (tracking pad) | _____ |
| Clearing debris disposal | _____ |
| Other: | |

_____

_____

# Negotiating the Price of Your Land

Sometimes the price of the land you want is at odds with your budget. If that's the case, it's time to put your Property Costs worksheet to work. The following sections cover some ways you can cut the cost of your lot.

### Compare Area Lots

If you're working with a real estate agent, he or she can help you find comparable lots and pinpoint their selling price. If you're flying solo, visit your town's tax assessor's office to get an idea of what land is going for these days.

### Don't Offer the Asking Price

Asking prices in real estate are usually like sticker prices on cars—a starting point. Unless the competition for land in your area is red-hot or the price on the lot is very low, consider the asking price a starting point and offer a lower bid. (And if the price on the lot is super-low, be sure to thoroughly check out the property to be sure the discount isn't to cover up some nasty problem with the land.)

After you make an offer, expect some back-and-forth negotiating. The seller can always counter, and you can meet somewhere in the middle. A good starting offer is 10 percent below the asking price. Remember, you can always negotiate up, but it's much more difficult to negotiate your price lower than your initial offer unless you find something seriously wrong with the property.

### Land Flaws Are Leverage

Sometimes property flaws or restrictions can be a good thing— if they can help you get the lot for a lower price. If the property will require heavy excavating, or if you're going to have to put in a septic system or well because public water and sewer are not available at the lot's location, be sure to point those issues out as reasons why the lot price should be reduced.

For instance, Gwen and Hank bought a lot that needed to have the road leading to it extended to meet building requirements. Gwen and Hank used that as leverage to get the price on the lot reduced, which more than covered the cost of extending the road.

### Negotiate a Quick-Closing Discount

If you're prequalified, find out from your lender how quickly the property can close. If you can close in 30 to 45 days—or, better yet, if you're paying cash for the lot and can close sooner—use that as a negotiating tool.

Consider this: every month the property sits in the seller's hands, he has to pay taxes. If the property has a mortgage, he has that expense, as well. If property taxes are $2,400 per year on the lot and his mortgage is $500 a month, for example, every month he holds the property is costing him $700. If you can close quickly, you're costing him less money than a buyer who needs 3 months or more to go through the approval process and close the loan, or than if he has to wait a few more months before another suitable buyer comes along.

Finally, the two golden rules of negotiating real estate are (1) do not fall in love with a particular piece of property and (2) be prepared to walk away if the deal isn't right. Keep your cool about the property you're trying to buy. When you become too emotionally invested in a specific piece of land, you're more likely to overpay for the lot.

## Dollar-Saving Do's and Don'ts

- Evaluate your lot carefully, and be sure you consider all the potential costs before you make an offer.
- Take advantage of for sale by owner (FSBO) or discounted builder's lots.
- Make any variance requirements the seller's responsibility.
- Do your homework to uncover and avoid any hidden costs and fees. Creating a Land Costs worksheet can help.
- Use the information you've gathered to negotiate the best price on your land.
- Ask for a quick-closing discount.

**DON'T TRIP** on your **SHOESTRINGS**

Don't sacrifice the quality of your location to save money. The old adage about having the worst house on the best block is true. Your location is essential for your home's overall value and resale potential. Choosing a lousy area to build a beautiful home will only ensure that you lose thousands of dollars in equity.

# 4

# DESIGNING YOUR DREAM HOME

Now the fun begins as you take the dreamy image you have in your head and begin turning it into reality. Whether you're having your building plans created from scratch or you find the layout of your dreams in a book of predesigned plans, you'll want to do some homework before you start building anything.

As you visit various homes, consider what you like and what you don't like about how the homes are laid out. Ask questions about what the owners like or what they would do differently. Note interesting features, convenience factors, and options that you'd like to add to your own home.

## Breaking Down Your Space

When you're designing your home, it helps to think about the space in broad categories to help decide how the space should be allocated. In general, the interior of your home falls into six categories: living space, eating and food-preparation area, sleeping room, work area, storage, and access area. Let's look at each.

### Living Space

These are the areas of your home where people gather and interact and may include a family room, den, living room, or media room. If you entertain frequently or if you have a large family who spends time together, ensure that these rooms are large enough to accommodate the groups of people who will be using them.

Think, too, about family pastimes. Does your clan spend lots of time watching TV and want lots of furniture to do so comfortably? Do you have large entertainment items such as a pool table or other game tables? If you're bibliophiles, you may want lots of space for books. Giving some thought to these issues now could save you disappointment later.

### Eating and Food-Preparation Space

The places where you prepare and eat meals include your kitchen, of course, and any dining spaces, such as a dining room, breakfast nook, or eat-in area in your kitchen. How many people do you need to accommodate in these spaces? Do you have a large family who needs to fit around a table? Do you really need a formal dining room, or will that space sit unused most of the time?

Also consider your food-preparation habits when planning your kitchen. If no one in your family likes to cook and you eat out often, it doesn't make a lot of sense to spend thousands on a big kitchen space with high-end appliances. However, if you believe the only reason you don't cook more is because you don't have the space or equipment, it makes more sense to create an environment more conducive to your culinary talents.

### Sleeping Space

Think about the areas where you'll catch your ZZZs. Does each family member need a separate bedroom, or will some share? Do you need a big master suite with a walk-in closet and full bath, or will a standard closet and half-bath do?

As you're planning your home, think about the future as well, especially if you're planning on being in this home for a number of years. With more adults dealing with "sandwich-generation" issues—caring for both aging parents and children—will you need a first-floor bedroom to accommodate an aging family member or individual with a disability? That will affect how your home is laid out.

Think about the positioning of bedrooms as well. Sleeping areas are better situated in the back of the house, away from street noise. If you have cathedral ceilings in your family room, consider the placement of bedrooms in relation to that opening to the second floor: if it's across from bedroom doors, you may find that the sound of activity in the downstairs area echoes loudly near the bedrooms, disrupting sleep. Try to consider any possible noise factors when you're thinking about where your bedrooms will be.

## Work Space

Work areas include such spaces as home offices, workshops, and utility rooms. They may be used for professional work or for home maintenance work, such as doing laundry.

For areas where you'll be doing professional work or hobbies that require a work space at home, think about the accommodations you need. Do you require high-speed Internet access and extra phone lines in that room? Do you need extra lighting or additional electrical outlets? Consider these issues in the planning stage and discuss them with your electrical contractor.

For areas that will be centers for housework, consider how to create the most functional space. For instance, does it make sense to add a sink in the utility room or garage for messy jobs, such as rinsing out mops, that you may not want to do in your kitchen or bathroom sinks? One woman we know found that her pet peeve when doing laundry was that the washer always finished sooner than the dryer, so her time wasn't used as efficiently as it could be. In her dream home, she designed a laundry room space with two dryers. She felt the expense of the extra dryer was worth it because of the time she saved in getting her laundry done. (Hey, it's her dream house, right?)

You should also consider where these work spaces best fit. A woodworking station might best fit in the basement, while an office for someone who works from home might be better in a room with a window and easier access to the home's entryway. Let function dictate how your home spaces are divided.

### Storage Space

Too little storage can end up being a headache, especially the longer you're in your home. Be sure to carefully consider how much storage your family needs. If you have young children, you need places to store clothing, toys, and the myriad other items children need and collect. If you tend to be a collector yourself, be sure you have places to stow your stuff. Otherwise, it'll end up as clutter around the house, or it might slowly take over an extra bedroom or your garage.

The construction of your home will also affect how much storage it has. If you opt for no basement, you may want to choose construction styles that optimize the space and access to your attic. After all, if the only storage you have is an attic, think about how you will get to that space. Will you need to teeter up a narrow drop staircase? That's not going to make storing furniture in your attic very easy.

If you have easy access to a basement, consider whether you can cut down on closet space to make the other areas of your home larger. For instance, if you have a full attic and basement, you may be able to keep off-season clothing in one of those areas, diminishing the need for a walk-in closet in one or more of the bedrooms. Larger living spaces mean you can cut down on the overall size of the house you need to build, saving money on your home construction.

### Access Areas

Access to your home includes the entryways, hallways, foyers, and landings. These areas connect your home—and, in the case of entries, offer the first impression.

As a rule, minimize these areas when possible. Although it's a good idea to have a foyer, a large entry area takes away from the functional space of the home. The same goes for hallways. Try to connect rooms from centralized areas to make the most of your living space.

Think carefully about the type of space you need and how you will allot it within your home. Some areas can serve double purposes—for example, an eat-in kitchen—so you might want to look for ways to combine them to reduce the space that you need in your home. For instance, you probably don't want to have your home office combined with your bedroom. However, it may fit nicely in combination with your guest room, which might only get used on weekends or during vacation times. This eliminates the need for two rooms and cuts your square footage by about 500 square feet.

## It's Gonna Cost You

Sure, it would be nice to have that 5,000-square-foot home with cathedral ceilings and hardwood flooring throughout. But every square foot you add to your home adds to your cost in several ways:

- **Additional materials.** When you add more features, walls, or other options to your home, it requires more lumber, drywall, insulation, paint, siding, and the like. Because you pay for all the materials, the addition comes out of your pocket.

- **Additional labor.** Extra rooms, closets, intricate framing, and odd shapes generally require additional labor, which will add to your cost. So although it might not seem like a big deal to raise standard 8-foot ceilings to 10-foot ceilings, consider this: in addition to the extra lumber, drywall, insulation, and paint, the new height might require the drywall installer to bring in scaffolding to reach the new height. That translates to more money.

- **Expensive options.** Granite countertops are beautiful, and Sub-Zero refrigerators are sure to spark envy from your neighbors, but you should ask yourself if such luxury touches are really where you want your money to go. Could the $10,000 that would pay for them be put toward a more efficient heating and cooling system or a better grade of windows that will ultimately save your family money on utility bills?

## Design for Your Property

As you design your home, consider some of the features of your new property, and consider those factors as you complete your design.

### View

If you have a terrific view, construct your house to take advantage of it. For instance, if your home faces water, you may want to move your master bedroom to the front of the house so you awake to the gorgeous view. Or if there's a picturesque farm to one side of your home, arrange your living space so the beautiful landscape complements your family room or living room.

### Slope

The slope of your property may affect the kind of home you can build. You might want to take advantage of the slope and incorporate it into the design of your home, adding a lower-level door to make a walk-out basement or a second-floor backyard patio. Or if your slope is in front of your house, consider putting your garage in the basement. You'll lose basement space, but you'll gain living space.

### Neighborhood

Your neighborhood should be a big consideration when you're designing your home, for a couple reasons. First, your home should fit into the general design of the area. If you're building a Spanish villa in the middle of a sea of Cape Cod homes, it's going to look out of place and you could very well have problems selling it later.

It's also a good idea to keep in mind the old adage about the worst house in the best neighborhood. That's not to say you should want to build a "bad" house, but you should opt for the best lot in the best town you can afford. This will add tremendously to the value of your home, because the homes in a so-called "good" town (good schools, good services, nice houses) typically hold their value and appreciate better than similar homes in towns without the same strengths.

## Making Decisions

As you begin the design process, check your desires and dreams against the realities of your lifestyle and budget. The decisions you make will affect how your home is constructed and appointed, so it's important to give a lot of thought to those expenditures that will really make a difference in your lifestyle.

### Consulting Your "Must-Have" Checklist

At this point, start looking at your "must-have" checklist you made in Chapter 1. Now that you've done some research, has the list changed? Don't worry—it probably has. It's not uncommon to make new decisions about what you want as you learn more about the options available to you.

The challenge comes when your "must haves" start exceeding your budget. That's when it's time to make some serious choices about the direction of your home design. In the upcoming chapters, we examine how to better understand the decisions that will impact the cost of building your home, as well as how to get comfortable with the compromises you're going to have to make.

### The Why Factor

As you look over your "must-have" list, ask yourself why you want that feature. Will it truly add to your quality of life? Or does the feature represent something else, such as making neighbors envious? If there's not a concrete reason for you to have it in your house, move it to the "nice-to-have" list. Ego has gotten more than one homeowner/builder into trouble.

### Dream vs. Reality

Again, look at the reality of your life. Don't go for the most expensive grade of carpet if you have three children and a slew of pets, because it's likely you're going to need to replace it in a few years. If you regularly entertain, design your home with a good deal of open space so guests can gather and not feel cramped. Spend some time thinking about how you spend your time indoors, and plan your home accordingly.

### The Waiting Game

Remember, you can always do some projects later. Crown moldings, flooring upgrades, and the like are nice, but you might want to get into your home and experience living there before you decide what makes sense for a splurge. For instance, the marble tile you've always dreamed of in your entryway could end up being a nightmare if your young children slip and slide on it every time it rains. Delaying some expensive upgrades can help you save the cost of ripping out an expensive option that doesn't work for your home.

## Designing to Reduce Future Costs

As you choose your materials, you can also keep in mind future savings by choosing the following:

- **Maintenance-free options.** Siding options such as vinyl and brick that don't require painting or expensive upkeep, or flooring options such as low-cost linoleum that can be wiped clean with a mop can be good choices for busy families who don't want to pay for or perform expensive upkeep.

- **Long-wearing options.** Some material options offer exceptionally long life, even if they require a bit of maintenance. For instance, wood floors generally require a bit more upkeep than linoleum, but when they're scratched or scuffed, they can be sanded down and refinished for far less than the cost of installing new flooring.

- **Climate-appropriate options.** Cedar shingles look great, but if you live near the beach, they'll likely become dingy and gray if you don't varnish them each year. Similarly, vinyl siding may be an inexpensive option, but if you live in a hurricane zone, don't skimp on the quality or you could find your siding lining the street instead of the side of your home after a big storm. Be sure you choose the materials most appropriate for your area.

## Common Design Flaws

As you design your home, keep in mind some common design mistakes. These may not seem like a big deal at first, but they'll begin to wear on you over time. Plus, they'll make a difference when it comes time to sell your home:

- **Poor flow.** Rooms that don't connect well or have insufficient openings between them can lead to poor traffic flow throughout the house. Open floor plans that use the minimum number of walls to support traditional framing are usually a good option and offer fewer opportunities for flow problems.

- **Cramped spaces.** Avoid creating small, dark rooms, which aren't really good for any purpose unless you're a photographer in need of a darkroom. Again, choose more open spaces with standard-size windows, if possible.

- **Poor indoor-outdoor access.** We know a family who thought eliminating the family room door to the backyard wasn't a big deal and thought it was a great way to get more wall space in the family room. Ultimately, they found it was a huge hassle to continually access their backyard through the side door in the garage, especially when they had friends over for barbeques. Don't make the same mistake—have easy access to your front and back yards.

- **Poor window placement.** Although oversize windows will cost you big bucks, you should invest in windows that allow in enough natural light as well as offer views of your yard, especially areas where children might play. Proper window placement will help you save on electric bills by providing more natural light and will give your house proper ventilation.

- **Illogical connections.** For obvious reasons, you wouldn't put a bathroom near your dining area or have a linen closet hidden in the back of one of the bedrooms. These inconvenient and unpleasant placements can spell real trouble at resale time, not to mention being annoying throughout the time you own your home. Be aware of room placement and how that will affect other rooms.

## Finding Free Ideas and Info

When it comes to designing your home, you have countless features and options available—if you can afford them. As you begin to get a better picture of your dream home, consult as many places from which you can get new ideas and information as you can. In addition to the magazines discussed in Chapter 1, try some of the ideas in the following sections.

### Builder's Models

Many new developments have model homes, which are usually outfitted with the most luxurious options available. These models are created as sales vehicles for the builder, so they usually have many upgrades and professionally decorated interiors.

Even if the model homes are out of your budget range, it's worth visiting them to get ideas and inspiration. You may get ideas that can be scaled down or otherwise modified to fit your budget. For instance, you might love a type of window in the model home. Although the model-home window might be a top-of-the-line, expensive brand, you can take that idea and shop for a similar style window in a less-expensive brand. You can also apply the same principle to home décor features.

Plus, by visiting actual structures, you can get a firsthand feel of how different home styles and floor plans flow. It's much easier to visualize how a floor plan will look in real life if you've been in a home that's similar. Take a camera along and take pictures, if you're permitted to do so. If not, jot down some notes and sketches as you find things you like.

### Brochures

From window sellers to home builders to building supply stores, you can find a wide variety of glossy, color brochures that can give you great ideas about the floor plan you'll choose.

### Websites

The Internet is a great resource for so many things—including ideas for your new home. Check out websites such as the National Association of Home Builders at www.nahb.org, especially the "For Consumers" section. Some of this section focuses on remodeling, but it also provides ideas on planning, which can be very helpful. Also www.reddawn.com has information about building environmentally friendly homes, as well as a bulletin board to reach other builders interested in doing the same. For a free demonstration of software that lets you draw your own plans (more on that later), check out www.dcad.com.

Or you can go to your favorite search engine such as Google or Yahoo! and type in "home design" or "building your own home." You'll find many sites you can peruse.

### Home Shows

Keep an eye out for local home shows, where various contractors and suppliers gather to display their wares and services. These shows often feature seminars about various home building and remodeling topics. Plus, if you take a list of questions to a contractor's booth, he or she will often give you free advice in the hope that you will ultimately hire him or her in the future. This is also a great place to grab some brochures.

**DON'T TRIP**
on your **SHOESTRINGS**

Collecting brochures, magazine clippings, notes, and photos is a good idea. But if they're scattered all over your house, you may lose track of all the good information and ideas you've gathered. To keep everything organized, create a simple binder sectioned off by room and filled with clear sheet protectors. File your visual aids in the appropriate section; then, when you sit down to plan your master bedroom, kitchen, or bathrooms, you'll have all your great ideas at your fingertips.

## MONEY IN YOUR POCKET

Take advantage of home show discounts. Some exhibitors offer a dollars-off or percentage discount if you book a job or place an order as a result of meeting them at a home show. Keep track of the vendors and contractors you meet at home shows, and cash in on these discounts where you can. If you get 10 percent off a $5,000 job, it's worth reminding the vendor where you met.

**Savings for You: $500**          **Running Total: $47,800**

### Home-Related Retailers

Home-supply stores and specialty retailers often have displays that can give you great information and ideas. Visit kitchen cabinet stores to see different sample kitchens, and check out bathroom displays in plumbing supply stores for great ideas. The professionals in these stores can also offer great advice, but be aware that their goal is to sell their products. Bring your questions and take the opportunity to pick their collective brains.

### Open Houses

You're not really shopping for an already-built house, but you can still visit real estate open houses as another way to get ideas about layout, design, and décor. Realtors may also offer virtual tours of houses on their websites. Check the real estate classified section in your local newspaper for both.

### Local Colleges

If your local college has an architecture, interior design, or construction management program, tap into the expertise there. Call the college and find out the names of instructors. They may be willing to discuss your ideas and point out potential pitfalls.

You have so many opportunities for free information and advice; don't spend a penny on professionals until you've exhausted your free resources. It's likely that you won't have to spend a penny to get the ideas you need.

## Dollar-Saving Do's and Don'ts

- Tap many sources for free ideas, advice, and information.
- Note what exterior and interior features may look great, but add thousands to your cost.
- Choose maintenance-free and long-wearing options to reduce your costs in the future.
- Design your home to maximize its resale potential.
- Consult your dream home checklist (from Chapter 1) to find out what your choices are. They may have changed, so don't consider them set in stone.
- Start working on your budget to determine what choices you'll need to make.

# 5

# SAVING ON YOUR STRUCTURE

Before you spend too much time looking at or creating your floor plan, you need to be aware of some of the exterior and interior features of a house and how they will affect your cost. You can make some decisions in this area that will save you money in the long run, as well as give you maximum flexibility as you make your selections.

## Saving on Your Exterior

Understanding how the outside of your home will impact your budget is important before you set your heart on a floor plan. You're bound to see some terrific-looking plans with lots of unique angles and features. Those plans are so unusual because they can add tremendously to your cost. Here are some exterior elements you should be aware of:

- **Odd angles.** Remember what Hank calls the "square rule." Typically, the more square or rectangular your home is, the more economical it is to build. Homes that have odd-shaped foundations, angles, or bump-outs will be more expensive to build because of the added labor and materials.

- **Single vs. multiple stories.** It costs more to excavate and build the foundation of a 1,900-square-foot ranch home than it does for a 2,000-square-foot Colonial. Why? The foundation of the same-square-footage home is twice as big when it's all on one floor. In other words, the foundation of a 2,000-square-foot single-story home is 2,000 square feet; the foundation of the same-size home in 2 stories is 1,000 square feet. The bigger the foundation, the bigger your cost.

- **Height.** Be sure the height of your house does not exceed the height restrictions in your municipality. If it does, your plan won't be approved. Also be sure you are clear on how the measurement is made—whether it's taken from the street to the top of the roof or from the foundation to the top of the roof. That will make a big difference in whether your plan is compliant. If your house is not within the height restriction, find out if you can modify the slope of the roof or reduce the height of the stories to the required height. Modifications cost money, but not as much as making the change after construction has begun.

- **Facade.** If the front of your home is ornate, with odd angles or expensive materials, you're adding to your budget. Look for ways to simplify to save cost. Fan designs in the siding or special brick ornamentation can be eliminated, for instance.

- **Roof.** The more simple the roof, the more cost-effective it is to build. Watch out for dormers, extra angles, and other fancy features that add cost.

- **Decorative windows and doors.** Floor-to-ceiling windows may look dramatic, but they add cost both in construction and in utility bills. Watch out for odd-shape or oversize windows, as well as "eyebrows" (half-circle or arched windows that sit above your traditional rectangular window) and doors with sidelights (slim windows that sit to one or both sides) and transoms (windows that sit over the top of the door)—all of which can be costly.

  Also be aware of window placement. Windows over tubs or in shower areas, as well as windows less than 18 inches from the ground, are generally required to be tempered glass, which will better withstand temperature changes. Tempered glass means more money.

- **Side-entrance garages.** Moving your garage doors to the side of your house may seem like a no-big-deal way to make the facade more attractive, but consider the additional paving you'll need to gain access to that side entrance. More paving means—you guessed it—more money. Plus, you'll ring up some additional costs on your front-facing siding.

- **Elaborate front porches and entrances.** That wraparound porch may be great for a rocking chair and a glass of iced tea, but it will increase your labor and cost. Also watch out for ornate porticos and entranceways. Balconies also add cost.

- **Skylights.** Any additional cuts to your roof and the installation of a skylight will add cost. Also skylights can leak if they're not sealed properly.

We're not saying that your best choice is a plain-vanilla rectangular box. You want your home to reflect your style and preferences. But be aware that, as a rule of thumb, the more elaborate the design, the more the cost. When you're watching your budget, you need to be aware of the factors that add cost and pick and choose the features that will make your home your own, while foregoing those that really don't make a difference.

## Interior Saving Tips

Interior features can also impact your budget and add cost where you might not realize. Before you pick a plan, consider how the following will affect your budget:

- **Open spaces.** Broad expanses of open space can mean your framing costs will increase. Vast expanses often call for expensive steel beams or other elaborate framing techniques. That means more money.

- **Vaulted or cathedral ceilings.** Raised ceilings don't necessarily add to the cost of building your home, although if they have curves or odd angles they may add to the cost of your wall materials and coverings. They do, however, minimize the functional space within your home (a cathedral ceiling means you lose that living space on the second floor) and can add to utility bills.

- **Bathroom basics.** Take a look at how many bathrooms you really need, because outfitting many baths adds to your cost in construction and may add to your property taxes. Consider how many fixtures you need in each bath: a half-bath with only a toilet and a sink is less costly than a full bathroom. Having a separate shower and bathtub instead of an all-in-one adds cost, as do features such as whirlpool tubs and multiple-head showers.

- **Electric.** Find out about the electrical code in your municipality. Some areas require that outlets be placed every 6 feet. If your plan has them placed every 8 feet, the plan will not be approved and you'll end up with expensive modifications. If the plan has more electrical outlets than are required, you'll incur more cost than necessary.

- **Plumbing.** Look for plans that have plumbing lines in clusters, such as bathrooms that are aligned on first and second floors, or bathrooms and laundry rooms that back up to each other. If you have long plumbing lines running through the house, you'll add additional cost in materials and labor.

- **Doors.** It's a pretty simple concept: The more doors you have, the more doors you're going to have to buy and install. Similarly, the more ornate or complex the doors, the higher the cost. Sweeping French doors are nice, but they're expensive. Look at areas where you might be able to replace a pricey door with something more affordable.

- **Complex or multiple staircases.** Spiral staircases, curved stairs, or simply multiple staircases in a home will add to your cost. See if there's a way to simplify the stair requirement.

- **Sunken features.** Adding a sunken feature can be dramatic, but you'll also add to the cost of labor and materials in your home. Sunken features also cut into the head room on the floor below, so a sunken family room over your basement may make part of that lower level much shorter than you would need to finish it as extra living space—something to keep in mind if you plan on using it in the future.

- **Arches and angles.** These types of features, as well as columns, shelving units, and others, are common between rooms and in entryways. They can add character, but they also add cost. Remember Hank's "square rule"—the more square or rectangular, the less expensive to build.

- **Fireplaces.** Obviously, the masonry work and framing required for a fireplace will add to your cost. In addition, if you're planning for a gas fireplace, you'll need to pay to have gas lines run.

- **Built-ins.** Anything built in—closets, shelving, nooks—adds framing, sheetrock, and finishing costs. Beware of lots of little spaces that need to be built in the construction phase—they can quickly add up to big bucks.

## Construction Considerations: Foundation

We're going to get into much greater detail about construction in later chapters, but before you choose your plan, you need to consider some construction basics.

The foundation of your home is the bottom-most part upon which the rest of the house sits. It's a good idea to find a skilled contractor to handle the excavation and construction of your foundation, because it's critical that this component, more than many others, be done well. If your foundation is uneven or sloped, it will cause problems throughout the rest of your home's construction.

Foundations start with footings, which are concrete pillars or mini-walls that extend into the soil underneath the home. The footings add stability to your structure and ensure that it can withstand the pressures of its environment. The depth requirement of your footings will depend on your area of the country.

On top of the footings sit the floor and walls, if any, of your foundation. The walls of your foundation should sit at least 18 inches above ground—or whatever your local building code requires—which prevents the wood structure from coming in contact with the ground, which could possibly cause rot or create a prime environment for termites. Financing through the Federal Housing Administration (FHA) generally requires that an 18-inch space exist between the bottom of the joist and the top of the basement floor for adequate access.

You can choose from several different types of foundations. Before you make up your mind, be sure to check with your municipality to find out if there are restrictions on the type of foundation you can choose. Otherwise, your options are listed in the following sections.

### Pier or Pile Foundations

Found mostly in areas near water, pier foundations, also called pile foundations, have wood or concrete "piers," or posts that extend into the home's concrete footing. A wood subfloor is then built on top of the piers.

This is generally the least-expensive type of foundation to build, but isn't permitted in all areas. It also provides no insulation to the bottom of the home, which could lead to greater heating bills.

### Slab Foundations

Much as you'd imagine, a slab foundation is a solid slab of concrete poured on top of the footings, usually with a layer of gravel between the earth and the concrete. The house is built directly on top of the slab, which then becomes the house's subfloor (the surface on which your floor coverings are laid).

### Crawl Spaces

A crawl space is a foundation that has a small gap between the footings and the flooring. The gap can range from 18 inches to a few feet, but it is not high enough to allow an adult to stand.

This type of foundation is less expensive than a basement and allows access to the area under the floor, which can be helpful for hiding and accessing heating and cooling systems, electrical work, and the like. Crawl spaces are often used in areas that are prone to flooding.

If you're going to eventually finish your basement and will need plumbing, you should decide that now. It's much less expensive to make accommodations for plumbing at the early construction phase than it is to add them later. You need to consider what plumbing fixtures you'll need, and whether you need any special pumping system if the level of the basement bathroom is below the level of the sewer or septic system.

### Basements

Basements, whether full or partial, add additional space for storage or finishing to the home. Basements can be entirely or partially below ground, and to save money, they can be combined with a partial crawl space or slab foundation.

Whatever type of basement you choose, consider how it will be constructed. Poured walls, where the concrete is literally poured into forms and allowed to set, costs more than traditional cinder block walls, but tends to be more waterproof.

In addition, consider the level of waterproofing you will need. The higher grade of waterproofing material you will need, the higher the cost.

## Framing and Flooring

As with foundations, you should take a few framing factors into consideration as you decide on a house plan:

- **Framing.** Big, open spaces may mean your framing plan needs steel beams. These beams can cost many times more than wood costs and may require that you rent a crane to lift them into place, which adds cost.

  Be aware of the materials your plan requires. In some cases, plans can be modified to include less-expensive materials, either by shortening the span of the open space or by including columns or posts in the design, but that can be unsightly in some cases, and the modification will add to your cost.

- **Trusses** A truss is a framed structure that distributes weight and eliminates the need for a bearing wall. If your roof needs to be trussed, it could drive up your framing cost significantly, because trusses require more lumber and labor than other methods of roof framing.
- **Chimneys.** In addition to the lines of your roof, check out the chimney requirements of your plan. If you have one or more chimneys that require masonry work, this will add to your cost. Simple box chimneys, which are framed and constructed out of wood like the rest of your home, cost less. Find out what your plan requires.

## Siding

You might not be thinking too much about siding at this point, but you need to decide in the planning stage if you're going to have any kind of brick siding on your home. If so, your foundation plan will need to have a brick shelf or support for the brick on all sides that will be sided with brick. If this isn't available on your plan, you will need to hire an architect to modify it. After you've poured the foundation and started construction, it's a big, costly job to add a brick shelf, so make this decision up front.

In any case, where you have found an existing floor plan to which changes may need to be made to save you cost, you should find out if alternative versions of the plan that already include your modifications are available. If the plan is already drawn, you can ask for modifications to be done as a condition of purchasing the plan. That's not possible in all cases, but the builder or architect may agree to it being part of the plan cost or doing it for a smaller additional fee.

# Savings Through Energy Efficiency

As you choose or create your home design, you should also be on the lookout for energy-efficient options and opportunities. Making your home energy-efficient not only helps the environment, it also saves you money on your utility bills. A small investment here may end up paying off for years to come.

The energy efficiency of your home can impact more than your utility bills. According to the U.S. government's EnergySTAR program, tighter home construction—minimizing gaps that can reduce efficiency by letting heated and cooled air escape—can reduce drafts, noise, and moisture, as well as improve indoor air quality by keeping out dust, pollen, car exhaust, and insects.

According to the EnergySTAR program, the energy efficiency of your home is determined by several factors, outlined in the following sections, which you should discuss with your home designer and contractor.

## Tighter Construction

Air leakage accounts for 25 to 40 percent of the energy used for heating and cooling in a typical home. Reduced air infiltration combined with proper ventilation not only reduces your energy bills, it can also improve your indoor air quality. Outdoor air that leaks indoors makes it difficult to maintain comfort and energy efficiency.

Readily available building materials such as house wraps, sealants, foams, and tapes reduce air infiltration. Builders can use these tools to seal cracks and gaps in framing along with hundreds of holes made for plumbing, mechanical equipment, and electrical wiring.

## Duct Size, Sealing, and Location

In typical American homes, heating and cooling ducts leak 20 to 30 percent of the air being forced through them, wasting money as warm or cool air escapes. To improve efficiency, duct systems should be properly sealed and verified by a field test to reduce any leaks. Duct tape, which is commonly used to seal ducts, does not adequately seal the joints, nor does it last very long. UL-listed tapes or duct mastic should be used to seal all joints and seams in the ductwork. (UL-listed means that the product has met the standards of Underwriter's Laboratories, one of the world's leading product safety and testing organizations.)

Home builders should consider duct size and location, too. Too-large or too-small ducts decrease the efficiency at which temperate air is transferred. And although builders often place ducts in out-of-the-way spaces such as attics, crawl spaces, garages, or unfinished basements, the extreme temperatures that can occur in these spaces (in the summer, attic air can reach above 150°F) can affect the temperature of the air moving through the ducts and into the home. Installing ducts within the conditioned area of a home will substantially reduce duct air losses.

In addition, to reduce these temperature variations, ducts need to be insulated. Ducts in temperature-controlled areas of the home need less insulation, but they should still have some insulation to ensure that the conditioned air is delivered at the desired temperature and to prevent condensation on the duct walls.

## Improved Insulation

Understanding insulation is simple: the colder your climate, the more insulation your house needs. In most climates, it's easy and cost-effective to increase insulation levels beyond those required by state building codes. This increase helps a home maintain a comfortable inside temperature while using less energy. For a home to maintain temperature efficiently, a continuous boundary of insulation is necessary between the inside and outside. Insulation must be installed carefully with no gaps, crimping, or compression, because these can allow unwanted air and heat exchange between the outside and inside. Pay careful attention in areas where the insulation has to fit around obstacles such as pipes, electrical wiring, and outlets.

### High-Performance Windows

According to EnergySTAR, windows typically comprise 10 to 25 percent of a home's exterior wall area and account for 25 to 50 percent of the heating and cooling needs, depending on the climate. Therefore, when constructing a new home, it's critical to choose windows that will be energy efficient. You should also ensure that windows are sealed around framing and any other gaps that may exist. Caulks, foams, and weather-stripping work well to keep out drafts, so be sure you consult with the contractor installing your windows to find out how he or she seals them. You may also want to be on-site to inspect the job and be sure the sealing process is completed as agreed upon.

In addition to better energy efficiency, high-performance windows can decrease noise in your home; reduce fading of curtains, furniture, and flooring (low emissivity coatings block up to 98 percent of UV rays); and be easier to operate than other varieties. (We cover how to choose the best windows in Chapter 7.)

Along with choosing good-quality windows, you'll want to size and position them wisely. To be more energy efficient, reduce the size or number of windows.

### Ceilings and Fans

Higher ceilings mean higher energy bills. So keep that in mind when you opt for vaulted or cathedral ceilings.

Also, plan for ceiling fan fixtures in your rooms. Ceiling fans are great tools to circulate both warm and cool air, saving you money in any season.

### Know Your Code

Your local building office will have energy-efficiency requirements for your area, such as the type of insulation you need to use and the like. Your local utility companies may offer free efficiency evaluations that can save you money on future bills.

Of course, you also want to choose energy-efficient heating and cooling systems, which can significantly reduce your utility bills. Such systems are becoming more affordable—and more efficient—every day. (We discuss them more in Chapter 8.)

Make your home more energy efficient now, and you'll save yourself headaches—and bigger bills—later. We talk about specific energy-efficient considerations in materials in the building chapters, but keep these general features in mind as you design.

---

## MONEY IN YOUR POCKET

The savings from energy-efficient design add up quickly. A savings of 20 percent on your gas and electric bills is possible. Calculate that on a combined utility bill of $200, and you have just earned more than 2 months of free utilities!

| | |
|---|---|
| **Savings for You: $480** | **Running Total: $48,280** |
| (*on a $200/month bill*) | |

---

# Dollar-Saving Do's and Don'ts

- Build up, not out, to save money on your foundation.
- Keep your design as rectangular as possible, and eliminate odd angles to save money on construction.

- Keep utility lines and units centralized to minimize plumbing and heating and cooling expenses.
- Use EnergySTAR guidelines to make your home more energy efficient and save you money for years to come.

# 6

# FINALIZING YOUR FLOOR PLAN

Your floor plan, also called a blueprint or house design, is the drawing that indicates the layout of your home and the specifications to which it will be built. You generally have two options for getting a finished floor plan: buy an existing plan, or create your own.

## Getting Started

Based on your budget, you should have made a few decisions on what you want in your house, including the following:

- Number of bedrooms
- Number of bathrooms
- Total square footage
- Number of stories

A good way to start getting an idea of your perfect house plan is to start sketching one on your own. Go through the binder of clippings, notes, and photographs you've been keeping, and start to cull out the features and room designs you love and would like to see in your home. Pay attention to the layout in homes that you visit. For instance, when Hank and Gwen started looking at floor plans, they knew they would occasionally have their combined (read: big) families over for dinners and celebrations. They wanted an open floor plan with a family room that opened to a deck or backyard. They also knew that they needed the dining room and living room adjacent to one another so there would always be room to extend the dining room table for another guest—even if it meant going into another room of the house!

You don't have to be an artist, and you certainly don't have to draw the sketch to scale, but by creating a rough layout, incorporating styles and features you like, you'll be better prepared to recognize the floor plan for your home when you find it. Or if you're working with a design professional (draftsperson or architect; see "Modifying Your Plans" later in this chapter) to have your plans created, he or she will better be able to see what you have in mind by looking at your sketches.

## A Word About Modular Homes and Home Kits

Before we move forward, we'd be remiss if we didn't mention the option of purchasing a modular home or home kit. Modular, or prefabricated, homes are literally delivered, already constructed, in pieces to your lot. They are then assembled into a finished home. These homes can be affordable, and the construction, if you purchase from a reputable company, can be quite good. However, most of these homes have limited options and floor plans, so you sacrifice the ability of being able to truly customize your home.

Home kits, such as those for building log homes, provide some of the materials and the step-by-step instructions for building your own home to the builder's specifications. Usually you will have to supplement the material in the kit with purchases from local suppliers. Again, this can be affordable, but you're limited in your options.

Two pitfalls to avoid in either case: first, be sure your lender will work with you if you choose this type of home. The supplier of the home or kit may want to be paid upon delivery, whereas your lender will likely want to see the finished home before giving a disbursement. Check with your lender and the supplier to find out what their requirements are. Usually, they'll both work with you, as long as they know about the situation in advance.

Second, be sure the home or kit meets all the building specifications for your area. If you find that the kit or structure is not compliant with local codes, you'll have a big headache and expense on your hands.

## Off-the-Shelf Floor Plans

Myriad books and websites offer existing floor plans in more styles than you can likely imagine. Bookstores carry books with hundreds or thousands of floor plans you can order from a clearing house or directly from the designer. In addition, a number of websites let you search for home plans that meet predetermined criteria such as square footage, number of bedrooms, number of bathrooms, or other factors. These plans usually range from a few hundred to a few thousand dollars, depending on the plan and the level of options you purchase. Considering that an architectural plan can easily cost you more than $10,000, predrawn plans represent a far more economical way to find a plan.

Even if you plan on having your home design created by a professional, it's a good idea to look through some of these home design resources to get more ideas and become familiar with how plans are drawn and look.

Here are some great websites to find floor plans:

| | |
|---|---|
| www.eplans.com | www.globalhouseplans.com |
| www.homeplanfinder.com | www.ubuildit.com |
| www.dreamhomesource.com | www.coolhouseplans.com |
| www.homeplans.com | www.houseplanstudio.com |
| www.areaplans.com | www.greathousedesign.com |
| www.10000homeplans.com | www.korel.com |
| www.dreamplans.com | www.bobvila.com |
| www.designbasics.com | |

Of course, there are many more, but you'll find more than 100,000 plans among these sites alone. That should get you started.

## Adapting Off-the-Shelf Plans

Another way to save a great deal of money on your plans is to find a plan that comes close to your ideal home design and then have an architect modify it for you. Architects can make simple changes, such as adding a closet or moving a drop-down stair from the attic, up to major restructuring and moving bearing walls within the structure. It's up to you.

We discussed the factors that will add cost to your home throughout the previous chapters. If you find a plan that you love but want to make some cost-adding changes, ask the company or individual selling the plans if they will make the changes or if they have alternative plans that reflect the modifications you want. Many of these plan services will modify plans for an additional cost. Get an estimate in writing, and compare that to what your architect or design professional will charge. Your design professional might make the changes for less money.

If you do intend to modify a plan, it's a good idea to speak to the architect about what you want to do before laying out money for the plan. Hank and Gwen faxed a rough sketch of a floor plan they found online to their architect with an explanation of the changes they wanted to make. The architect then told them what was possible—and what wasn't—and offered ideas to improve the plan. In our case, this review was done free of charge, but some architects may charge a nominal fee for this kind of review. However, this is a good investment. Had we bought the plans and not been able to modify them to our specifications, we likely would have abandoned the plans, wasting the $800 we spent to purchase them.

If you find a set of plans that need many modifications, though, those changes could be just as costly as having new plans drawn. It may pay to take note of what you like about those plans and keep looking for a layout that will incorporate those features, but better suit your needs.

## Reproducing Plans

You're going to need a number of sets of plans throughout the building process. If you buy stock plans, it's a good idea to pay a bit more for the reproducible plans. Because house plans are copyrighted by the professional or firm that created them, you can't make copies of the plans without permission, and fines for copyright violation are $10,000 per occurrence! Get permission to reproduce your plans. Plan options also include so-called vellums, overlays which enable you to make changes to stock plans.

Some plans even come on a CD, and you can modify the plans with a CADD (computer-aided drafting and design) program, either your professional's or your own.

**DON'T TRIP**
on your **SHOESTRINGS**

Your municipality may require that off-the-shelf house plans be certified by an architect or engineer licensed in the state where you intend to build. Call your municipality's land use or building office to find out what the requirements are for getting approvals on existing plans.

How many sets will you need? Here's a good estimate:

- You
- Lender
- Land use office
- Excavator
- Footing/foundation contractor
- Framing contractor
- Plumber
- Electrician
- Heating, ventilating, and air-conditioning (HVAC) provider

That's nine copies right there. You may be able to re-use some of the plans. For instance, when your excavator is done with his copy, you may be able to get them back and let the framing contractor use the same set of plans. But this doesn't always happen as expected, so it's better to have a few extra sets of plans to avoid delays.

## MONEY IN YOUR POCKET

When you do make copies of your plans, get all your plans done at once, rather than in batches as you need them. Per-page printing costs usually go down as you increase quantity, so if you buy 10 sets at $10 each once, you're paying less than if you bought 5 sets at $15 per set twice. That's a cool $50 savings.

**Savings for You: $50**　　　　　　　**Running Total: $48,330**

## Designing Your Own Plans: Floor Plan Software

There's no rule that says you have to have someone else draw your floor plans. If off-the-shelf isn't for you and you're certain you want something that's uniquely your own, you can design your own plans. As with shopping for an off-the-shelf design, you'll first want to start with rough sketches of what you want in your home and the home's layout.

To help you turn your rough sketches into plans, explore home design software. Many packages enable you to customize home designs and require varying degrees of skill to operate. Some packages even give you warnings if you've made a mistake—if the span between walls is too broad, for instance. Other great features include estimating tools and materials lists. With prices generally about $30 to $300, they can also help save you money.

Although we don't endorse any particular products, here are some you might want to compare:

- *Better Homes and Gardens Home Designer Suite 6* (www. homedesignsoftware.com)
- *myHouse 3D Home Design* by DesignSoft (www.designsoftware.com)
- *Punch! Professional Home Design Packages* (www. punchsoftware.com)

Be sure the package you choose matches the level of computer and design skills you have. If it's not clear, contact the company's customer service department and ask a few questions before purchasing.

## Hiring a Design Professional to Modify Your Plans

You may have noticed that we've been referring to a "design professional" up until now and not necessarily always "architect." You do have a couple options when it comes to choosing someone to modify your plans: a draftsperson or an architect.

When we say "design-your-own" house plans, we don't mean that literally. It's best to work with a professional to ensure that the plans are structurally sound and that the design is appropriate and livable.

### Draftsperson

A draftsperson is a skilled professional who can complete a set of scaled drawings for you based on your needs and indications. Many drafting professionals work with CADD software packages and help the designer render floor plans as well as plans for plumbing, electrical, and other systems. Working with a draftsperson can potentially save you thousands of dollars over working with an architect.

To find a draftsperson, look in your local phone book. Sometimes these professionals will advertise in the service directories found in many local newspapers. Or if your local college has a drafting or CADD curriculum, ask for referrals there.

### Architect

An architect is an individual who is highly educated in the field of home design, holds a degree in architecture, has completed an intern- or apprenticeship, and is licensed by the state in which he or she practices. Of course, with all those credentials comes a hefty price, and having an architect design your plans can cost you anywhere from $10,000 up, depending on the complexity of the plans, the difficulty of the area in which you're building, and other factors.

The benefit of having an architect draw your plans is that they will be drawn to the code requirements in the municipality in which you want to build. In addition, if you make minor changes to the plans as you progress through the construction process, the architect will likely do them for free or for a modest fee.

Hank and Gwen opted for off-the-shelf plans and had them modified by an architect. In addition to the modifications we made, the architect suggested various other changes we hadn't

**DON'T TRIP**
on your **SHOESTRINGS**

Before you choose one design professional over another, be sure you know what your state and municipality require. Different states have different plan certification requirements, so even if you save money by hiring a draftsperson, you may end up having to hire an architect or engineer to certify the plans.

thought of. For instance, he noticed that our plans had drop-down stairs leading to the attic at the top of the stairs on the second floor. He thought that would be unsightly, so he moved them to an area of the second floor that was more out of the way but still allowed us easy access to the attic. Our architect also repeatedly checked in on the home construction and was available to answer questions from our subcontractor.

## Finding an Architect

Architects are licensed by the state and usually belong to one of the professional societies, such as the American Institute of Architects (AIA; www.aia.org). You can probably find one fairly easily. However, the best means of finding an architect are usually referrals and recommendations:

- Ask friends and neighbors, especially if they've used an architect for design work you like.
- Speak with local builders and interior designers.
- Call your state licensing board.
- Track down your local AIA chapter and ask for recommendations.
- Check the phone book as a last resort.

You may want to find out if the recommended architect has a website. If he or she does, check it out to get an idea whether the architect specializes in commercial or residential work and a glimpse of the architect's style. This can also give you clues about the architect's price range, which will vary from professional to professional. If most of the homes featured on the website are elaborate mansions in the most affluent areas, this probably isn't an architect who will negotiate with you on price or who will be particularly sensitive to the budget-conscious builder.

## MONEY IN YOUR POCKET

If you want to save money on an architect, go back to school—your local college that is. If you live near a school that offers an architecture program, approach a student to draw your plans for you. The student will get much-needed experience, and you'll get cheaper plans. An architect could easily charge more than $10,000 for a full set of plans; an architecture student may work on the plans for half that amount. It's a good idea to specify that you want a senior or someone who's apprenticing, and ask the dean or a professor for someone they think would do a good job. If your municipality requires it, you can get the plans certified, or "stamped," by a professional engineer or architect, again for much less than creating custom plans.

**Savings for You: $5,000**          **Running Total: $53,330**

You have a few names of architects in hand, and you're ready to begin the process. What's the best way to narrow down your choice and get started on the right foot?

### Make Some Calls

Phone the architects on your short list. While on the phone, spend a bit of time describing your project and your needs to find out if the architect does the type of work you have in mind. Ask if the architect has done projects similar in size and scope and if he or she has worked in the area in which you want to build. If you get a good feel on the phone, set up a time to meet and see samples of the architect's work. You might also ask for addresses of homes the architect has designed so you can drive by and check them out.

### Ponder the Past

When you meet with the architect, look at his past projects and ask questions. Does he answer you to your satisfaction? How does he describe previous clients? Does he characterize many

of them as difficult or unreasonable? Does he describe the projects as challenges? If so, that may be a red flag, and you'll want
to check his references closely to be sure the problem was, in
fact, the clients and not him.

### Listen Closely
As you explain your project, is she really listening to you and
coming up with answers to your questions and concerns? You
don't want someone who is going to force his or her opinions
on your project.

### Check Out the Surroundings
An architect who does business in a dark, dingy basement raises
the same concerns as a hair stylist who has a bad dye job. Many
fine architects are based out of their home or small offices. In
fact, because these professionals have lower overhead than big
firms, they may be a better choice for your budget. However,
their surroundings should still indicate a sense of professionalism and should reflect some of their architectural style. Look for
indications of professional credentials and affiliations, such as
licensing and society memberships. If these are not clearly visible, ask about them. Again, if the basic tenets of professionalism
aren't clearly evident, check references closely.

### Answer the Architect's Questions
Expect the architect to ask you questions about your lifestyle.
The architect will want to know how you spend your time in
your house and what's important to you. And if the architect
isn't asking these types of questions, that could signal a problem.
This helps the architect understand what type of house will
be best suited to your family. Bring along your pictures and
sketches to help illustrate what you have in mind.

### Confirm Where the Architect Has Worked

Ask whether the architect has done work in the municipality in which you're building. It's important that he be up to date on codes and requirements in that area. Plus, having an architect who knows who's who at town hall can be a great resource if you run into problems with your building inspections later on.

It's also helpful to know that your architect has worked on projects similar to yours. Even if you discussed this on the phone, ask to see photos of houses that are somewhat the same construction as yours. They don't need to be exactly the same, but it's helpful to know that this architect has experience with the same size and style of home that you want to build.

### Bring Up Money

Don't be shy about talking about your budgetary issues early in the conversation. Architects can be wonderful resources for finding cost-saving measures—they've seen firsthand most of what works and what doesn't. Your architect can point out, for example, where sacrifices can be made in framing or where you can make materials substitutions. When Hank and Gwen met with their architect, he pointed out that the house plans called for brick on three sides, which was going to be very expensive. By scaling back the plan to include brick on the facade only, we saved on the foundation cost as well as the siding cost.

Some architects may also have predrawn plans for sale that would fit your needs—and your budget—so be sure to ask.

While you're talking about money, ask about the architect's rates and how you will be charged. Most architects either charge an hourly rate or tack on a percentage of the overall cost of the project. Find out how long she expects the process to take and if there are any up-front charges you need to consider. Find out if the plan fee includes everything you need to build the house,

**DON'T TRIP**
on your **SHOESTRINGS**

If your architect isn't sensitive to your budget or doesn't seem to understand that you need to be careful about costs, thank her for her time and leave. Plenty of other architects will work with you to build a house within your means.

including electrical and plumbing plans. Get a written estimate of the architect's services and costs.

## Request References

Ask the architect for references. Contact the references and find out how involved the architect was in the process. You might want to ask some of the following questions:

- Did the architect listen to your input?
- Was the architect easy to reach with questions and concerns?
- How long did it take to get the plans?
- Were any revisions required? If so, how long did it take for the revisions to be done?
- Were there any unexpected charges?
- How did contractors react to the plans? Were the plans clear and complete?
- Did the architect make him- or herself available to contractors if they had questions?
- Did the architect check in on the construction to see that it was going as planned?
- Were you happy with the end result?

## Meet and Greet

You'll need to meet several times with your architect to be sure the plans develop as you want them to. Be sure you and the architect have ample time to go over the plans and make modifications. Be sure you understand the plans, which can be complex.

## Stay Involved

Negotiate with your architect up front that he or she will be involved throughout the building process. Some architects will want to conduct site inspections throughout the process to ensure that the plans are being executed correctly. Ask your architect to give you a separate quote on making on-site inspections. This could be a very worthwhile investment, as he or she will know, often at a glance, whether the job is being done properly.

## Plans You'll Need

A full set of architecture plans has several different components, so when your architect quotes you a price, be sure it includes the following:

- **Key.** This section explains the building data—a summary of the pertinent information about your house, including the number of stories, height of structure, building area footprint, square footage, and the like, as well as an explanation of symbols used and other details that will help you understand the plans.

- **Specifications.** This is an explanation of how the work on each section of the house has to be completed, including details about the construction of the foundation, framing, and even the nail sizes used.

- **Framing plan.** This portion of the plan includes the floor framing plan, exterior framing, interior framing, and sheathing specifications. It indicates the placement of beams and the type of framing that's to be used.

- **Cross sections.** These illustrations show the interior of the house at various points, as if you had sliced open the house and were looking inside. It helps you see how your rooms align from floor to floor.

- **Footing and foundation specifications.** This section of the plan shows the foundation's layout, dimensions, materials, and other information your excavator and mason will need to properly build your foundation.

- **Elevations.** These are the various views of the house's exterior—front, back, and each side.

- **Plot plan.** This is the scaled overview of your property and where your home and other structures are going to be placed. It marks setbacks, easements, septic systems, decks, pools, and other building locations and parameters of the property.

- **Floor plans.** Simply, these are the plans that show the room layout on each floor of your house. They include room sizes, window dimensions, and framing specifications. You can find sample floor plans in Appendix C.

Your plan set may also include the following:

- Electrical plan
- Plumbing plan
- Heating and cooling plan
- Materials list
- Details about trim and cabinetry specifications
- Window and door specifications (interior and exterior)

# GO FIGURE

Compare the pricing of various plan design options and see which fits your needs and your budget.

| Option | Estimate |
| --- | --- |
| Custom plans from architect | _____ |
| Drawn own plans: | |
|  Purchase software program | _____ |
|  Work with a draftsperson | _____ |
| Use stock plans | _____ |
| Use architect or draftsperson to<br> customize plans | _____ |

As you can see, creating a plan is a pretty hefty order. Covering all the myriad details that contractors need to do the job well takes time and skill. Before you tackle the project yourself, we recommend consulting a professional, especially if you're a first-time homebuilder.

## MONEY IN YOUR POCKET

Whether you purchase your plans from an architect or from a plan designer, it's not necessary to invest in the materials list. The suppliers from whom you purchase your materials will often create these lists for free so they get your business. Materials lists run about $100 or more, so that's a nice chunk of change to save. Be sure, however, to be meticulous about checking the window estimate to be sure the supplier didn't miss anything. It's also a good idea to run the framing list by your framing contractor for input before you place your order.

**Savings for You: $100**          **Running Total: $53,430**

## Getting on the Same Page

When you've selected your architect, you'll want to set a budget up front and get your agreement in writing. (See Appendix C for a sample contractor agreement.) Your agreement should include the following:

- The scope of the project, with a clear description of the work that will be completed by the architect
- The timeline, or how long it will take to complete the plans
- The list of deliverables, including exactly which plans and specifications are included in the project
- The fee schedule, including when payments are due

- The number of plans with the architect's seal (necessary for your town and, possibly, for your lender) you'll receive

- Permission for you to reproduce the plans as needed or a clear fee schedule to receive additional copies of the floor plan

- A guarantee that the plans will meet the code requirements of the town in which you're building or will be revised without additional cost

- An agreement that the architect will be available if questions arise during construction

- A provision that you will be notified before any additional charges are incurred (so you have no nasty surprises when you get the final bill)

- A clear indication of the cost and turnaround time for future changes to the project

Be sure any and all agreements, deliverables, and expectations you have are outlined in the contract, no matter how small. Remember, your contract is the document to which you will refer if something is not delivered as you expect. If it's clear and complete, it's less likely you'll run into any misunderstandings about everyone's role.

## Working With Your Architect

You'll be working closely with your architect for the weeks (or longer) it will take to design your house plans. To make the relationship go more smoothly, produce the best possible results, and cost less, keep the following rules of thumb in mind:

- **Be prepared.** Time is money in most cases, but especially so when you're dealing with your architect. By the time you sit down with your architect, you should have a good idea of what you want. Bring any sketches or photos of features you want in your house. Being as specific as possible about what you want will save on the architect's time—and your ultimate cost.

- **Set a schedule.** You should have regular meeting times or conference calls with your architect to check progress and go over changes.

- **Be clear.** Don't say things such as "I don't like it, but I don't know why" or "I'll know it when I see it." Your architect is not a mind reader. Be specific about what you like or don't like about the plans. For instance, if you're unhappy with the facade of the house, try to determine whether it's the look of the siding, the size of the shutters or the windows, or the symmetry. Even if you can't name your issue exactly, explain what bothers you as clearly as you can.

- **Be realistic.** Your architect is working for you, but don't assume he's at your beck and call. And although your architect should be accessible, understand that sometimes he may need to get back to you instead of taking your call immediately. As long as you're getting good service, try to be patient.

- **Group changes.** After your plans are completed, try to only make the changes you think are necessary. Also try to submit any changes you do have in groups, because your architect will likely charge you less to complete a few changes submitted together than those submitted one by one. It usually takes less time for him to make a few adjustments at one time than it does for him to finish one, put away your project, then take it out and work on it again.

The more specific you are, the less time you take and the less money you have to spend. Architects usually bill based on the amount of time a project takes, so you'll cut your costs by being clear and concise.

## Planning Within Your Budget

Whether you're designing the home yourself or working with a design professional, keep the following points in mind to help you cut costs in the final design of your plan. (These are in addition to using lower-cost materials, which we discuss in later chapters.)

### Keep It Simple

Simplify elaborate plans, such as changing expensive windows to less-expensive models, replacing special-order sizes with standard sizes, or making the floor plan more rectangular.

### HUD Helps

Follow the principles developed by the U.S. Department of Housing and Urban Development (HUD) and National Homebuilder's Association (NAHB) Research Foundation. These guidelines, which we discuss more in later chapters, were developed to give homebuilders some guidelines for building their homes more cost-effectively. The full set of guidelines is available for $150 at www.toolbase.org. The site, sponsored by the NAHB, HUD, and others, also has a lot of great free information about economical homebuilding.

### What's the Weather?

Consider area-specific requirements for your home. If you're building in a cold climate or an area that's at risk for hurricanes, be sure you have a good understanding of what measures to take to make your home more energy efficient, durable, and able to withstand the climate. This will save on energy and, possibly, repair bills later on.

## Move It Up

Position your home closer to the street on the plot plan. This will decrease the length of the driveway and save on driveway paving costs, as well as the cost of running pipes, wires, and other utility hook-ups, which get more expensive with the distance at which they need to be installed.

## Streamline Inside

Ditch some walls and hallways to create a more open floor plan, which will make your home appear larger inside. This will save on framing, drywall, and interior insulation costs. Instead of building a wall between the kitchen and family room, keep the space open, with just a half wall or small shelving unit to divide the space, if necessary. Instead of creating a long hallway, consider how you can expand your rooms to reduce that space.

In a home that Gwen and Hank remodeled, the original design had a hallway that spanned the length of the house, with a bedroom entry on the right. By expanding that room into the hallway area and adding the bedroom entry straight ahead at the end of the hallway, we added about 39 inches to the width of the bedroom, which made it much more functional.

## Finish Later

Cut down on expensive trim and decorative features—for now. Don't pay for a molding plan or include plans for elaborate trim and finishing aspects around your house. Keep it simple; you can always add things such as crown molding later on.

## Build Up, Not Out

Remember that a 2,000-foot, 2-story home is cheaper to build than a 2,000-foot, 1-story home because the latter's foundation is much larger. You'll also generally incur more in plumbing, heating, cooling, and other charges because more lines and ducts need to be run throughout a 1-story house than a 2 story house, where utilities can be kept centralized.

## Keep Sizes Standard

Extra-high ceilings require additional lumber, sheet rock, and labor, adding to your overall cost.

Also remember that as your home gets bigger, your prices rise in more ways than one. The 5,000-square-foot estate you're planning may require a sturdier foundation and framing as well as far more insulation, wallboard, finishing, painting, and carpeting.

Your structure may require additional footings, or specialized framing to support the weight of a larger home. Although you want to build a home that will be spacious enough to accommodate you and your family now and in the future, budget-minded builders should be conservative when planning their square footage.

**DON'T TRIP**
on your **SHOESTRINGS**

You can consult an area real estate agent or the land use office in your municipality to get an estimate of the cost per square foot to build a home in your area. For instance, in some areas, you might find that a home costs an estimated $125 per square foot to build. We don't really put much stock in those numbers. The cost per square foot of your home depends on too many variable factors, including the structure, the materials used, who does the work, etc. We don't recommend spending too much time or energy trying to figure out what other people paid per square foot. Instead, apply the budget-boosting tips in this book to build your home at a bargain price.

## Remember Resale Value

Even if you think this is the last house you'll ever own, you need to keep resale value in mind as you're designing. It's very dangerous to create quirky design features that appeal to you but make no sense in good home planning.

For example, at one time Gwen and Hank were home shopping and came across a Cape Cod that had an oddly shaped, three-story addition. The owner-builder had constructed the addition to have a family room on the bottom floor, with a workshop sectioned off by a half wall. The living room was then a loft above those two rooms, so every time someone was watching television or working on a project downstairs, the noise carried easily to the living room, which should have been more peaceful. With a second-floor wood fireplace (all that hauling wood up the stairs) and two bedrooms that were too tiny to be functional in the design, it was ultimately too odd for us to make the investment, even for the reasonable price at which it was offered.

The home will need to be sold someday—whether it's by you or your heirs. If your home is too oddly designed, it could also affect your ability to get an equity loan, which would limit your ability to leverage your assets to pay for a college education or other investment.

Consider some features that always enhance resale value:

- **Curb appeal.** How will your home look from the outside? Is it attractive and of a style similar to other homes in the area? Or does it look out of place for the area? Does it have features that make it look unbalanced? (Symmetrical facades are often better than asymmetrical.)

- **Convenience counts.** Have you designed the home with common convenience features and expected amenities? For instance, does it have ...

    - A separate laundry room?

    - A linen closet near the bedrooms?

    - Adequate counter and storage space in the kitchen?

    - More than one bathroom?

    - Hidden drop-stairs or attic access?

    - Appropriate heating and cooling features? (Most new homes today are built with central air conditioning, except in the coldest climates.)

## Working With Your Checklist

We've given you many considerations to ponder as you develop your home design. As you work through all of them, be sure to consult the checklist you filled out in Chapter 1. You'll find that many features commonly considered "need-to-haves" are, in fact, features that will enhance resale value. In the upcoming chapters, we discuss how you can save money on your overall structure to fit in other "nice-to-haves." (Granite countertops, anyone?)

However, even as we give you tips to make your home more affordable and efficient, be sure the decisions you make are ones you can live with. It's important to be prudent in your decisions, but if you have a great view from the front of your home and want to enjoy that view from your master bedroom, don't just dismiss that idea because you'll hear more street traffic. Weigh your options given your unique situation, and make the best decision for your family, your budget, and your circumstances.

## Dollar-Saving Do's and Don'ts

- Thousands and thousands of floor plan designs are available at much lower cost than custom-designed plans. Review them for ideas. You might even find one you love.

- Home design software makes it easier to truly design your own plans, often at an even lower cost than off-the-shelf plans.

- If you already have an idea of what you want, turn to an architect, architecture student, or draftsperson to create your plans for you.

- Be frank with your home designer about your budget, and he or she can suggest more affordable home designs and materials.

- Move your home closer to the street to save on paving and utility installation costs.

- Design your home with resale potential in mind to maximize your investment.

# 7

# READY, SET, BUILD!

You've got a great plan for the home of your dreams. Now you need to understand the processes that will turn your two-dimensional blueprint into your life-size dream home.

Many steps and considerations go into the process of building a home. Whether you're working with a general contractor (GC) or overseeing the project yourself, it's important to use these first steps to lay the foundation of your project, both in a literal sense—getting the property ready and building the structure that will support your home—and in a figurative sense—where you begin managing the project in a way that will yield the best possible results. You'll have a better handle on whether the project is staying on track or not, and you'll be better able to spot areas in which you can save time and money.

It can't be stressed enough: know what inspections are required before you begin the building process. Moving on to the next phase of a project before the previous phase has passed inspection will usually end up costing you big bucks.

# Preparing Your Lot

After you've gotten your permits and approvals to build, the first visible step to beginning your project is to get your lot ready for what's to come. That will happen in three primary phases, and all will likely be done by your excavating contractor.

Your excavator will need a copy of the plot plan your surveyor prepared. Different municipalities require different certifications from your surveyor; sometimes you won't be able to get your permits until your clearing limits are set. In Gwen and Hank's case, the survey was conducted by the engineering firm that reviewed their plans. It included the following:

- **Property survey.** This means that the property boundaries were identified for the purpose of the survey. It's often necessary to transfer title.

- **Plot plan.** This is the overhead view of your lot, including where all the structures will be placed. It is necessary for the building permit.

In addition, Gwen and Hank's agreement with the surveyor included him returning to the lot at the appropriate times during construction to …

- **Set clearing limits.** These are the boundaries within which you are permitted to clear trees and brush. If you clear beyond these limits, you may find yourself facing stiff fines.

- **Stake the building location.** This marks where the home will be located on the property and shows the excavator where to dig.

  At this point, you should request that the surveyor also set a benchmark for the finished floor elevation. This is a point visibly marked, perhaps on a tree or stationary post nearby (be sure it can't be moved!) that shows the point where the first floor will align. This helps the excavator know exactly how far down to dig, taking into account measurements for flooring, supports, sills, and the like. If these measurements are off, your building could end up being over the height requirement and require significant cost to correct.

- **Certify the foundation location.** This means the foundation is officially complete and is in the right place. Framing can now begin!
- **Complete the final survey.** This final survey after the building is complete certifies that the structure is positioned properly and meets township specifications for size and height.

Always check with your municipality to see what it requires of the survey process, and negotiate those steps into your fee. It's a good idea to choose a surveyor who is familiar with the requirements of your municipality. While you're checking on the survey requirements at your town or city's building office, it may be a good idea to ask for recommendations of surveyors.

## Clearing the Lot

Your lot will need to be cleared of any obstructions to building. Trees need to be knocked down or cut; stumps and large roots need to be dug out; any structures, large rocks, or other obstacles to building need to be removed. This will likely be done by the excavator, but you could also hire a subcontractor who specializes in harvesting and selling wood to do the work. Often, local firewood vendors will come in and either cut the trees in exchange for the wood or pay you a small fee for the wood, depending on the type that's on your property. This will eliminate the charge as part of your excavating bill and may reduce the fee the excavator will charge you for disposal of debris from the property—and earn you some additional cash. In many cases, your excavator can simply put the cut wood aside until it can be removed. Find firewood vendors in your Yellow Pages or in your local newspaper.

During the clearing, pay close attention to your plot plan. You may be required to keep trees and brush around the perimeter of your property for the width of the setbacks or more. If you clear too far, you could be fined hundreds or thousands of dollars, or be required to replant the area, which will also cost you.

Don't be quick to clear all the way to your clearing limits, however. Wooded areas can provide privacy, and mature trees are very desirable. In addition, having a barrier of mature trees around your home could cut wind, potentially saving you money on your utility bills. Plus, keeping mature trees is a good environmental move. You can always clear more trees later, but it will take years to grow back overcleared property, so be conservative about how much wooded area you remove.

### Grading and Soil Work

While you're clearing your lot and digging your foundation, you also need to consider the approved grades the property needs to meet. Your excavator will likely be responsible for leveling the property, if needed, or creating slight slopes so water runs off appropriately to avoid having it pool in your yard or flood your basement.

Your excavator should strip off the uppermost layer of soil—called topsoil—from your property and put it aside. Topsoil is the darker, richer soil that will feed the landscaping on your property. Under the topsoil is the more sandy, less organically rich soil, which is what should be manipulated during the grading process. After your lot has been graded, the topsoil can be replaced, creating an ideal habitat for grass, plants, shrubs, and other landscaping to grow.

Failure to strip off the topsoil and replace it means it will likely be covered or mixed in with other grades of soil. If this happens, you'll need to purchase additional topsoil to cover your property; otherwise, you won't be able to grow a blade of grass. Save yourself later expense by saving your topsoil now.

### Creating a Tracking Pad

A tracking pad is like a driveway or access path covered with gravel or stone. Many large machines—cement trucks, delivery trucks, cranes, bulldozers, etc.—will need to be on your lot throughout the building process. You can avoid wear and tear on the soil and the other areas of your property by having your excavator put down some gravel to clearly mark where trucks should enter and exit.

### Digging

After the lot has been cleared, the stumps and other obstacles hauled away, and the topsoil stripped, it's time to start the big dig! Breaking ground on your home is a very exciting step and is usually the point when all this planning and thinking starts to become very real. We recommend that you be on-site when your foundation is dug, not just because it's a great moment to witness, but to be sure the dig doesn't turn into disaster.

It's critical that you or your contractor (or both) check and double-check that your foundation is marked correctly in both size and position on the property. Before one shovelful of dirt is dug, measure the markings that outline the dig area. Also measure from the edges of the foundation markings to the edges of the property. If your foundation is dug in the wrong place, especially if it doesn't meet your setback requirements, you'll end up with a huge headache and additional expense. In the worst case, you'll need to have your foundation dug again. In the best case, you'll be required to apply for a variance, which will cost you both time and money. Remember, check twice (or more), and dig once!

We strongly recommend having a professional dig your foundation unless you are very knowledgeable in using excavating equipment. Professional excavators can create very level grades in the bottom of your foundation, requiring less work to correct unevenness. Think carefully before you skimp on any part of the foundation of your home.

### Supplying Contractor Needs

As the work on your lot progresses and your contractors are spending more time on-site, you'll need to make provisions for some of their needs. They may need electrical access for equipment, which can be provided through a generator they bring with them or by your contacting the electric company to initiate your service. They may also need a portable potty and some running water. Discuss these needs with your contractors and how they generally work so you're ready for them. Happy contractors are usually better contractors.

## Laying Your Foundation (and Basement)

We discussed your foundation options (see Chapter 5). Now it's time to take a closer look at how foundations are created. This is the underpinning of your home, and it should be well constructed, or it could lead to myriad problems later. The following sections list some things you should think about when working with your foundation.

### Footings

Footings are cement fixtures upon which your house sits. They anchor your home to the ground and widen the base of your foundation, giving it additional stability and distributing the weight of the house. The size of your footings are largely determined by the size of your house, the slope and stability of the ground upon which it's being built, and the local building code requirements. Footings need to be poured within the perimeter of the house as supports for posts, and additional footings may be needed for heavy house features, such as chimneys.

When pouring your footings, you need to consider the frost line (the depth at which frost penetrates the earth) in your area. It can extend down as far as 4 feet in cold climates. Of course, the deeper your foundation, the higher your cost. However, the National Association of Home Builders Research Center (NAHBRC) outlines a frost-protected shallow foundation (FPSF), which basically insulates the structure and allows it to be built above the frost line, in many cases. Thermal insulation protects the footings and allows them to be 12 to 15 inches below grade, according to the NAHBRC. This could potentially save you thousands of dollars, so check with your architect or with your municipality's building office to see if it's a money-saving option for your home.

### Slab

The slab is the horizontal plane of your foundation. It is a layer of concrete that sits on top of your footings and a layer of gravel. If you've opted for a basement, you may want to consider delaying the pouring of the slab, which would be the floor of your basement, until your basement walls are framed. That way, the slab can dry protected and you're less likely to have marks from debris landing in the wet concrete. Of course, with quick-drying concrete, this is less of a problem, but it's still something to consider. You will also need a slab poured as the floor of your garage.

If you have a basement, the area under the garage should not be excavated. The garage floor should sit on solid ground, as the weight of cars and/or equipment is too much to be supported by beams.

### Basement Walls

If you decide to include a full basement as part of your building plan, you need to consider how the walls will be constructed. We discussed briefly the difference between cinder block walls and poured concrete walls (see Chapter 5).

Poured concrete is more expensive, but it can offer better protection against water entering your basement and give you more control over the quality of the wall. Poured concrete walls are constructed by creating forms, or molds, for the walls. Rebar, or steel rods, are placed within the forms and embedded in the concrete to reinforce it.

Many building codes don't allow wood foundations, usually built from pressure-treated lumber, but they can still be found in some places. We generally don't recommend them because the wood has a greater tendency to decay than concrete.

At the top of your walls, you need to make some accommodation for connecting the walls to the frame of your home. This is generally done with a J-bolt or strap anchored into the concrete. Check your local building codes for requirements.

Keep in mind that if you're going to have a wood fireplace, you need to know that up front, because you need to have a fire box—a masonry box that protects your home's wood framing from the heat generated by the fireplace. You also need to plan for a chimney and flue, which will vent the heat and smoke generated by the fireplace up and out the roof of your home. If your home will be using natural gas or propane, you can opt for a fireplace that ignites using that fuel, instead of wood, which will not require a fire box or chimney and will save you money.

### Waterproofing

Waterproofing material is generally sprayed on or painted around the base of the foundation to keep out water. Crawl spaces also generally require a moisture barrier on the floor of the foundation, usually made of plastic. Different waterproofing materials have different guarantees. Synthetic materials are often guaranteed for life.

Also consider whether you'll need a foundation drain or sump pump if you have a basement. Both will help you remove water from your basement in case of a flood, which can happen due to environmental water entering the basement, a pipe bursting, or other water-system malfunction. It also may be necessary to meet the building code in your area, so check that out.

### Backfilling

After your footings or basement walls are set and waterproofed, your contractor will take some of the dirt that he dug out to create your foundation and use it to fill in the gap around your foundation. (Be sure topsoil isn't used for the backfill—you'll need that later.) At least 18 inches of foundation should still appear above the ground, with any appropriate cut-outs for basement windows and the like.

### Saving Money on Your Basement

You may be able to save money on your basement walls by opting for precast concrete foundation panels. The NAHB's ToolBase Services (ToolBase.org) describes precast panels as steel-reinforced concrete studs, with reinforced top and bottom beams and concrete facings. The cost is estimated at about $45 per linear foot, which puts it at about the same materials cost as poured concrete. However, because the walls are estimated to go up in about 4 to 5 hours and no on-site concrete work is required, you can expect to save at least several hundred dollars on the work that

Your municipality may require that an inspection of the foundation be done before it's backfilled. Be sure an overzealous excavator doesn't backfill before the inspection is complete, or you'll find yourself having to dig out the foundation again so the inspection can be completed.

has to be done. They can also be ordered with cutouts for doors, windows, and ledges.

You'll want to discuss the cost of delivery with the professionals who will be doing the installation. Sometimes a crane needs to be rented to hoist the panels into place. However, that may also be included in the installation estimate you receive, so be sure to check. Also check with your municipality to ensure that the panels meet the code standards for your area.

## Sticks on Stones: Framing

When the framing of your home starts, that's when you really begin to see your home come to life. Framing consists of the interior "sticks"—the 2×3, 2×4, or 2×6 boards that create the support for your interior and exterior walls, flooring systems and sheathing, and the exterior panels of your home. Your framing contractor may also be responsible for installing your windows, but you need to negotiate that up front.

### Types of Framing

You should understand the different approaches to framing so you can discuss your best option with your framing contractor and your material supplier.

- **Platform framing.** Sometimes called conventional framing, platform framing literally means that the home is built one floor at a time, with each floor acting as a "platform" for the next. This is the more common—and more economical—form of framing found today.

- **Balloon framing.** This type of framing uses solid, extra-long beams from the sill of the foundation, all the way to the roof. If your home is two stories, your vertical beams are two stories high. This is a very solid way to build a home, but, of course, the extra-long lumber is pricey and requires additional labor to maneuver, so it's more expensive. However, it's easy to run electrical and plumbing lines through the walls because there are no barriers. Check with the building office and fire inspection office on code issues related to balloon framing.

- **Post-and-beam framing.** The Canadian Wood Council describes post-and-beam framing as using long, thick structural members to provide support. The beams are the structure, exterior sheathing and interior finish support, as well as the insulation cavity.

### Sills

Your framing starts at the foundation level, with sills— usually planks of pressure-treated wood that are adjoined to the foundation using J-bolts or straps. A metal plate or piece of pressure-treated lumber, called a sill plate, may be between the foundation wall and the sill. Your contractor should also use a foam called a sill sealer between the wood and foundation. It's important that sills be affixed tightly and securely to your foundation to keep the rest of your framing secure.

The sill is the first part of your framing to be installed. Before the sill is installed, it's very important to check to be sure the foundation is level, or flat. An uneven foundation—even one that's off by an inch or so—can cause problems throughout the building process, so you want to be sure the foundation is as flat as possible. You can check this by using a tool called a level, which has a clear cylinder marked with two lines near the middle. Within the cylinder is liquid with a small bubble. When the level sits on a flat surface, the bubble rises squarely between the two lines. If the bubble sits to one side or the other of those lines, your surface is not level and needs to be corrected before framing continues.

Do *not* assume that the mason checked the foundation or that the framing contractor will. Check it yourself just to be sure.

After your sill has been installed, your home will likely have a main support beam called a girder. This may be steel, or it may be crafted from a number of planks nailed or glued together. It usually sits in notches in your foundation and is supported by posts or by the foundation directly.

The next steps depend on the type of framing being used for your house, but some things are constant, as the following sections outline.

### Flooring

Your floor joists, the beams that support the floor, are connected to the girder in your home. Here, you have a choice of choosing dimensional lumber, which is the garden-variety 2×12 wood beams that come in different lengths—or engineered flooring systems. Dimensional lumber flooring systems include joists that connect to the girder with tongue-and-groove plywood nailed on top to create the subfloor (the platform on which flooring material is installed).

Engineered flooring systems are manufactured beams, usually crafted from wood particles and a bonding material. Because manufactured beams don't expand and contract like traditional lumber, the creaking noises that many wood structures have aren't an issue with engineered wood products. They're typically more straight and more rigid than lumber as well.

### Wall Construction

Interior and exterior walls are created by nailing a series of 2×4 (or 2×6) studs approximately 16 inches from center to center of stud, to a top plate (basically, studs placed horizontally at the top and bottom—think about it like a ladder on its side).

Usually, walls are built as independent structures and then raised and attached to the home's framework. As your walls take shape, you'll see the skeleton of your home and begin to see how the layout looks, how big the rooms really are, and how your home is coming together.

When you look at framing construction, it's important to ensure that the walls are plumb, or perfectly vertical. Nonplumb walls will cause problems when you install windows and drywall. Walls that tilt outward are not structurally sound, either. Again, your contractor will probably check for this, but you should, too, using a level.

### Sheathing

Think of sheathing as your home's skin. It's generally plywood, although some builders opt for manufactured boards, called pressboard. The sheathing is nailed to the studs that frame your home, and after the sheathing is in place, your home will really begin to look like the exterior elevations you've seen in your plan drawings.

**DON'T TRIP** on your **SHOESTRINGS**

You may see some tract builders using pressboard for sheathing. Pressboard is a manufactured board that looks like plywood, but is really smaller shards of wood glued together. Pressboard is cheaper, but we don't recommend using it. Wet pressboard warps and loses its stability, so if the board is exposed to wet weather before it's enclosed by a vapor barrier and siding or if, later on, a window leaks and water comes in contact with the sheathing, you could find that your wall is warping, which will affect your siding.

We strongly recommend spending the extra money for plywood. Check your building codes and see what the minimum plywood requirement is, because you may be able to save money by choosing a code-minimum thickness of plywood instead of a thicker variety.

## Saving Money on Your Framing

You can work with your framing contractor to save money on this phase of building. Be sure you ...

- Have the right materials ready when the contractor starts. Delays cost money.

- Discuss the materials and framing techniques the contractor will use. See if there's a way to save on labor.

- Investigate some of the new high-performance beam products instead of using steel, if your plan calls for it.

- Use the right-size materials to reduce waste.

- Have materials delivered precut to framing plan specifications to cut labor costs.

You may also want to discuss the following money-saving ideas with your framing contractor and/or architect. Some of these are suggestions from the NAHB Optimum Value Engineering standards, which were created with the goal of setting more economical housing construction standards:

- Eliminate the sill plate, if your code allows.

- Reduce or eliminate the band joist or bridging between floor joists.

- Use smaller studs in interior walls. We generally recommend 2×4 studs, but you may be able to use 2×3 studs in some places.

- Eliminate headers on non–load-bearing walls.

- Use single lumber headers or top plates when possible.

- Use pre-engineered trusses to create tray ceilings for less cost than framing them.

# Windows and Doors

Windows are an important part of how your house looks, both from the inside and the outside. Windows let light come into your home, provide a way for you to connect visually with the outside world, and give you a way to ventilate your home. Energy-efficient windows can also help you save on energy bills.

Choosing the right windows—and the windows that fit your budget—can be a mind-boggling process. You'll find many different types of windows with many different types of features. To make the decisions about what each feature is worth to you, you need to understand windows and their components.

## Types of Windows

Different houses and even different rooms within the same house will require different types of windows. Windows come in a number of common styles:

- **Double-hung.** This is the most common type of window and probably what you think of when you hear the word *window*. A double-hung window has two sections that move vertically up and down, independent of each other, and you can open the top or bottom portion of the window. Single-hung windows have a fixed window at the top, and the bottom window moves up and down.

- **Casement.** This is typically a single window that opens out from the house with a hinge on one side of the window. The window usually opens with a crank mechanism.

- **Awning.** These windows are similar to casement windows but open away from the house with a hinge at the top. They use a crank action, similar to casement windows.

- **Bay/bow.** A bay or bow window is usually a series of three or more windows that are attached and angled out from the room in an arch. These are generally decorative windows used in living rooms or to showcase a specific view.

- **Slider.** A slider is like a double-hung window placed on its side. It's got two sections that move from side to side and don't extend away from the house. You'll often see sliders used as basement windows.

- **Eyebrows, arches, or transoms.** These windows can be arched, half circles, or rectangular, and are usually used on top of other windows as accents. They are lovely and decorative, but often very pricey.

You might also have a good reason to have fixed glass in your home—a window that doesn't open. These are usually used near the top of cathedral ceilings, where the window is inaccessible.

### Getting the Grade

Windows come in various grades and, as you would expect, the better the grade, the more expensive the window. With windows, you have lots of options, including the following:

- **Clear glass.** This is, obviously, the preferred material and is used in the majority of windows. It can be tinted to reduce glare, and new glazing options can increase heat retention.

- **Low-emissivity (Low-E) glass.** Surface coatings on Low-E glass help it retain up to 40 percent more heat without affecting the light that passes through the window.

- **Tempered glass.** Up to five times stronger than regular glass, tempered glass is usually used for large glass areas, such as glass inserts in doors or sliding glass doors.

- **Insulating glass.** Insulating glass cuts heat loss by using two or more panes of glass to enclose a sealed air space. This creates an additional barrier that reduces the amount of heat lost. Most windows today have at least two panes of glass.

- **Other features.** Many windows, especially double-hung, will have features that make it easier to clean them, including a tilt-in feature that lets you easily reach every part of the window. Elaborate hardware and finishing can also add cost.

Wood windows cost more than vinyl windows. In fact, according to the Vinyl Institute, the vinyl siding's trade association, one 3×4-foot wood double-hung window costs about 10 percent more than its vinyl counterpart, raising the bill for new windows by more than $650 for an average home, which they estimate at having 15 windows.

### Understanding U-Value

According to the Efficient Windows Collaborative (EWC; www.efficientwindows.org), a website that provides information about the energy efficiency of windows and is sponsored by the U.S. Department of Energy's Windows and Glazings Program, a window's heat loss is indicated in terms of its U-factor (or U-value). The lower the U-value, the greater a window's resistance to heat flow and the better its insulating value. The EWC recommends that northern-climate residents select windows with a U-factor of 0.35 or less; central-climate residents should opt for windows with a U-factor of 0.40 or less; and southern-climate residents will find windows with a U-factor lower than 0.75 and preferably lower than 0.60 helpful to regulate interior temperature.

### Window Ills

When your windows are installed, there are a few common mistakes you'll want to avoid. First, be sure the windows are level. If your windows are not level, it's glaringly obvious and will reduce your curb appeal.

Also be sure the framed opening for each window is appropriate. You don't want your windows jammed into a too-small space because you're likely to have problems opening your windows later. Wood windows may expand and contract with the weather, and a window in a too-small space may expand and jam.

Conversely, too-large of spaces can lead to drafts around the windows and decreased energy efficiency. Be sure your windows are shimmed (held in place by small pieces of wood that keep them tightly in place) and sealed with insulation before the drywall is installed.

### At the Door

When you shop for your exterior doors, you may get a bit of sticker shock. Elaborate wood door systems with sidelights and transoms can run as much as $20,000! Of course, that kind of expense will quickly break your budget, and you have much more affordable options.

Exterior entry doors are generally 36 inches wide by 80 inches tall and come in several material options, which will affect your cost:

- **Wood.** Wood doors can be flush or paneled. Flush doors are flat doors that are either solid wood or are constructed of two wood surfaces affixed to a core of another material. Wood doors run from moderately expensive to very expensive, depending on the type of wood and how ornate the door is. Wood doors can warp, must be maintained, and need to be refinished every few years.

- **Steel.** Steel is a very affordable and durable door option. It's maintenance-free and can cost far less than wood doors. They're generally very energy efficient.

- **Fiberglass.** Fiberglass doors have the same benefits as steel—energy efficiency and no maintenance. However, they also can be finished to look like wood and are easily repaired in case of damage. They're more expensive than steel, but they offer a richer look, especially if you choose wood-grain options.

Glass inserts, transoms (rectangular overhead insets that match the door), and sidelights (rectangular vertical insets that are installed on one or both sides of the door) will all add to your cost—sometimes a transom or sidelight is more than the door itself! Be aware of how quickly these ornate touches can skyrocket your door costs.

### Saving Money on Your Windows and Doors

You can save money on your windows and doors a number of ways:

- **Shop around.** Windows with the same features vary greatly in price from brand to brand and store to store. You could easily knock 20 to 30 percent off your price by comparison shopping.

- **Stay with standard.** Don't opt for expensive hardware or glass options, such as glass insets in doors or stained glass in windows. Choose standard hardware and save.

- **Choose less-elaborate windows.** If you don't need the one-button tilt access, forego it. You'll feel it in your wallet.

- **Frame 'em.** Negotiate with your framing contractor to install your windows and doors. This will often be quicker and cheaper than if you hire a specific window installation specialist. Your framing contractor may give you a much better price as part of the larger job.

Check the building department in your town to find out what the code requirements are for window sizes. Bedrooms may require windows large enough for an adult to escape in case of a fire. Install the wrong-size windows, and you could find yourself starting from scratch—at great expense.

- **Get the best of both worlds.** Choose windows with maintenance-free vinyl exteriors and wood interiors—cheaper than all-wood windows. Choose maintenance-free doors.

- **Snap in.** Choose snap-in grilles or caning—those insets that make the glass in your windows and doors look like smaller panes—which can be cheaper and easier to clean than permanent ones.

- **Finish last.** Order your windows prefinished from the manufacturer. At a few dollars each, it costs less than hiring a professional to paint them. Doors, on the other hand, are generally easy to finish, so save money by doing that yourself.

## Siding

When it comes to the siding on your home, you have a number of options, each of which will have a different impact on the look, energy efficiency, and cost of your home. Some of the most popular siding options today include those discussed in the following sections.

### Brick

Brick on the exterior of your home can create a beautiful look. Brick comes in far more varieties today than the traditional dark red; you can find brick in browns, reds, weathered effects, and many other different looks. According to the Brick Industry Association, brick adds up to 6 percent of value to your home. It's maintenance-free, termite-proof, and offers additional weather protection. When brick is secured with a quality and well-mixed mortar, it may not need any attention, other than the occasional power wash to clean it, for decades.

More popular today than solid brick is brick veneer. Brick veneer is simply a layer of brick and mortar laid over the sheathing of your home, creating the same effect as solid brick, but at much less cost. Still, this option isn't as inexpensive as, say, vinyl siding. Some homeowners opt for brick on the front of their homes and use vinyl siding or another option on the sides and back, or they simply use brick as an accent, such as on the exterior of a chimney. This gives the rich look of brick at far less expense.

## Vinyl

Made from polyvinyl chloride (PVC) plastic, vinyl is probably the most economical choice for siding and has largely replaced aluminum siding as an affordable external home covering. According to the Vinyl Institute, a 1996 study found that, taking into account installation costs and maintenance for 20 years, vinyl siding costs are 51 percent of cedar textured plywood siding, 64 percent of aluminum siding, and 37 percent of brick.

Vinyl siding comes in long strips that overlap and are nailed to the sheathing. Today, vinyl siding comes in a variety of shapes and many, many colors. Some is even molded to look like cedar shingle. Vinyl is also often combined with more expensive forms of siding. Homeowners can create an expensive facade and then use vinyl siding on the sides and back of the home.

## Cedar Shake

Cedar shake are wood shingles made of cedar. They look terrific when they're first installed, but the wood can soon discolor from weather and climate. Cedar shake requires a great deal of maintenance, including annual treatment. If you don't mind your shingles turning brown or gray—and many homeowners don't—you can let them weather on their own, but they still need a good varnishing now and then to protect them from the elements. Still, they don't require as much maintenance as wood clapboard.

### Wood Clapboard

Wood clapboard is a siding constructed of finished planks of wood that are affixed horizontally to the sheathing. The most popular choices are usually cedar, pine, spruce, or fir. Redwood is also an option, but it is very expensive. Wood clapboard requires regular maintenance, which may include paint or stain, depending on the finish you prefer. After a number of coats of paint, the clapboards must be refinished.

Different woods are more readily available in different regions, so you should investigate the wood that's the best priced and suited for where you live. Wood clapboard looks terrific, but it's relatively expensive and it's definitely not the best choice for those who are looking for a maintenance-free option.

### Stone

Stone gives a rustic, traditional look to a home's exterior. It's durable and has many of the same properties as brick. Common exterior siding stones include limestone, slate, and sandstone, among others. The variety of stone you see will likely vary depending on where you live. Stone available locally is probably far less expensive than a stone that would need to be imported from other areas.

Just as you can opt for brick veneer, you can also choose precast stone veneers, which offer the look and feel of stone at a much more palatable price. You can also opt to use stone accents—our friends opted for a stone accent on the bottom third of their traditional Colonial home facade. The stone gives the house a great look, and using it as an accent was far less expensive than using it on the full facade.

### Stucco

Stucco is a plaster made of cement mixed with sand and lime. It's applied wet to exterior walls or surfaces to create a continuous surface. Stucco has a classic feel to it, and can be finished smooth, rough, or patterned. Stucco has good insulation value and is less expensive than brick, but it's not as inexpensive as vinyl siding.

Stucco is very versatile and comes in many colors. More recently, synthetic stuccos have appeared on the market. These are less expensive, but they may not wear as well, especially if they're not applied correctly or they're used in harsh climates, such as in the Northeast or in seaside communities where the salt air can cause damage.

Whatever siding you choose, the basic process of installing it is often similar. First, a vapor barrier is affixed to the exterior sheathing. This thin sheet of material is used as a barrier against moisture, so it's usually plastic, but it could also be a builder's felt or other material, depending on the climate. Over the vapor barrier, a thin layer of rigid foam insulation is affixed to the sheathing.

The siding is then installed in the appropriate manner: vinyl panels, wood shingles, or planks are affixed; brick, stucco, or veneers are applied.

## Saving Money on Your Siding

For some money-saving siding tips, try the following:

- Use expensive siding as an accent rather than an entire exterior treatment.
- Shop around for the best price on materials.
- Use single-layer panel sheathing. Some siding products can be applied directly to the studs, with only a vapor barrier beneath.
- Limit trim, shutters, and other external décor.

# Up on the Rooftop

Your roof will shelter your home—and you!—from the elements for years to come. A number of considerations come into play when building your roof, and you need to understand those that will affect your cost.

The National Roofing Contractors Association (NRCA) says that all steep-slope roof systems—those with slopes of 25 percent or more—have five components:

- **Roof covering.** Shingles, tile, slate, or metal and an underlying material, such as tar paper, that protect the sheathing from weather

- **Sheathing.** Boards or sheet material fastened to roof rafters to cover a house or building

- **Roof structure.** Rafters and trusses constructed to support the sheathing

- **Flashing.** Sheet metal or other material installed in a roof system's various joints and valleys to prevent water seepage

- **Drainage.** A roof system's design features, such as shape, slope, and layout that affect its ability to shed water

### Types of Roofing

When selecting your roof system, you have a few considerations. Of course, cost and durability head the list, but aesthetics and architectural style are important, too. The products you choose will impact the cost of your roof.

The following roofing products commonly are used for steep-slope structures:

- **Asphalt shingles.** This is the most common roofing material in the United States. Steep-slope roofs can be reinforced with organic or fiberglass materials. Although asphalt shingles reinforced with organic felts have been around much longer, fiberglass-reinforced products now dominate the market.

Regardless of their reinforcing type and appearance, the physical characteristics of asphalt shingles vary significantly. When installing asphalt shingles, NRCA recommends using shingles that comply with American Society for Testing and Materials (ASTM) standards—ASTM D 225 for organic shingles and ASTM D 3462 for fiberglass shingles. These standards govern the composition and physical properties of asphalt shingles; not all asphalt shingles on the market comply with these standards. If a shingle product complies with one of these standards, it is typically noted in the manufacturer's product literature and on the package wrapper.

- **Wood shingles and shakes.** Made from cedar, redwood, southern pine, and other woods, the natural look of wood shingles and shakes is popular in California, the Northwest, and parts of the Midwest. Some local building codes limit the use of wood shingles and shakes because of concerns about fire resistance. Many wood shingles and shakes only have Class C fire ratings or no ratings at all. However, certain wood shingle products incorporate a factory-applied, fire-resistant treatment and hold a Class A fire rating.

- **Tile.** Clay or concrete tile is a durable roofing material. Mission and Spanish-style round-topped tiles are used widely in the Southwest and Florida, and flat styles also are available to create French and English looks. Tile is available in a variety of colors and finishes. Tile is heavy! If you're replacing another type of roof system with tile, verify that the structure can support the load.

- **Slate.** Available in different colors and grades, depending on its origin, slate is considered virtually indestructible. It is, however, more expensive than other roofing materials. In addition, its application requires special skill and experience.

- **Metal.** Primarily thought of as a low-slope roofing material, metal is a roofing alternative for home and building owners with steep-slope roofs. You'll find two types of metal roofing products: panels and shingles. Numerous metal panel shapes and configurations exist. Metal shingles typically are intended to simulate traditional roof coverings, such as wood shakes, shingles, and tile. Apart from metal roofing's longevity, metal shingles are relatively lightweight, have a greater resistance to adverse weather, and can be aesthetically pleasing. Some have Class A fire ratings as well.

- **Synthetic roofing products.** Also available today are various simulated traditional roof coverings, such as slate and wood shingles and shakes. However, they do not necessarily have the same properties as the materials they're meant to resemble. Synthetics are often less expensive and usually require less maintenance than their traditional counterparts.

Before making a buying decision, the NRCA recommends that you look at full-size samples of the product, as well as manufacturers' brochures. It also is a good idea to visit a building that's roofed with the product you're thinking of using on your home.

### Gutters and Downspouts

Gutters and downspouts help channel water from the top of your roof down and away from your home to prevent flooding and ground saturation, which can cause undue wear and tear on your foundation. Gutters are affixed horizontally along your fascia boards and channel water to downspouts, which channel the water vertically and then into a tray called a splash block.

Gutters and downspouts are available in galvanized steel (which is usually the least expensive) or aluminum that comes either finished or unfinished. Finished aluminum, which comes in a variety of colors, doesn't need to be painted, as the other materials do, before being affixed to your home. You can also save money by choosing a lighter-weight version of the metal.

Gutters and downspouts are affixed to your roof with metal pieces called strap hangers, for gutters, and downspout hangers for downspouts.

In theory, it's pretty easy to install gutters and downspouts, but you need to be comfortable working atop a ladder.

## Go Ahead and Vent

Don't discount the importance of ventilation in your roof. Improper ventilation can lead to heat and moisture buildup in the attic. That, in turn, can cause your rafters and sheathing to rot, and that results in buckling shingles and decreased energy efficiency. Not a good thing, no matter which way you look at it.

Louvers, ridge vents, or soffit vents all offer ventilation. Proper attic ventilation, including attic fans, will also help prevent structural damage caused by moisture, increase roofing material life, reduce energy consumption, and enhance the comfort level of the rooms below the attic.

In addition to the free flow of air, insulation plays a key role in proper attic ventilation. The NRCA says an ideal attic has ...

- A gap-free layer of insulation on the attic floor to protect the house below from heat gain or loss.
- A vapor barrier under the insulation and next to the ceiling to stop moisture from rising into the attic.

- Enough open, vented spaces to allow air to pass in and out freely.
- A minimum of 1 inch between the insulation and roof sheathing.

The general ventilation formula is based on the length and width of the attic. NRCA recommends a minimum of 1 square foot of free-vent area for each 150 square feet of attic floor—with vents placed proportionately at the eaves (e.g., soffits) and at or near the ridge.

Check your local building codes for attic ventilation requirements in your area. Your architect or home designer may also take into consideration such factors as sun exposure, shade, and humidity levels of your property and location.

### Soffits and Fascias

Soffits and fascias need to be covered, which your siding contractor will usually take care of. They can be covered with siding or wrapped in aluminum to match your siding material. The latter offers terrific protection against the elements.

### Saving Money on Your Roofing

Consider the following tips to save money on your roofing:

- Have the roofing shingles delivered to the site after the sheathing is put on the roof. In many cases, you can have the shingles placed on the rooftop, saving some labor costs.
- Look for discontinued roofing materials. Buy a bit extra for repairs later on.
- Eliminate overhangs on your roof where possible.
- Choose code minimum sheathing materials.

Now you have a working knowledge of all the steps—and all the potential savings—involved with building the structure and exterior of your home. These first steps are very labor-intensive and are critically important for the longevity of your home and the quality of life you'll enjoy in it. Now let's go inside.

## Dollar-Saving Do's and Don'ts

- Pay close attention to survey requirements so you're not hit with huge charges for correcting mistakes.

- Investigate with your building department and mason to determine whether a frost-protected shallow foundation is a money-saving option for your home.

- Stick with standard-size windows and doors to save money.

- Use expensive siding, such as brick and stone, as an accent rather than a full facade.

- Sell the lumber cleared from your lot to offset disposal charges for debris.

# 8

# ALL SYSTEMS GO

The systems in your home keep it warm and cool, provide water, power your electrical appliances, and keep everything functioning as it should. If the framing and exterior of your home are the backbone and skin, the systems are the internal operating mechanisms that keep your house alive. Understanding the different systems in your home will help you choose budget-minded options.

It is possible to do some of the system work yourself, but you'll want to check with your municipality before you plan that into your budget. Some local building codes require that a licensed plumber, electrician, or heating, ventilating, and air-conditioning (HVAC) professional handle the work.

## Water, Water Everywhere

Your plumbing system will deliver water into your home through pipes and return waste water to your sewer or septic system. Good plumbing systems help you conserve water and energy, with considerations for delivering hot water more efficiently.

You usually need a licensed plumber to install, or at least oversee and approve, plumbing for new construction, especially when it involves tapping into public water and sewer systems.

The plumbing system is made up of two subsystems, the supply system and the drainage-waste-vent system.

### Supply System

The supply system, also called your intake system, delivers water to your house through a large pipe called a main. The main to your house is connected either to the public water supply or your well, depending on where you'll be getting your water. This main then runs to the cold water inlet, with a shutoff valve in place just before the inlet. (The shutoff valve enables you to, obviously, shut off the water supply to your house.) Then, pipes snake off from the supply, delivering some water to the water heater and also throughout the walls in your home, ending at various fixtures and appliances as needed.

Because the water is delivered by pressure—the pounds of water being forced through the pipes—pipes can run both vertically and horizontally. Your plumber should ensure that your water pressure isn't exceptionally high. Pressure more than 80 pounds per square inch (psi) should have a pressure-reduction valve to regulate it so it doesn't cause undue wear and tear on your pipes.

### Drainage-Waste-Vent System

The drainage-waste-vent system (DWV) removes waste and water from the house and returns it to the municipal sewer system or septic system. Drainage pipes from various appliances and fixtures enter into a main drain. Because this system relies on gravity rather than pressure, to move waste water and material through the pipes, the pipes must slope down at least ¼ inch per foot of pipe or as per building code.

Gases build up in the DWV system, so the system has to be vented to allow gases to escape. A buildup of gases and, therefore, pressure, in a pipe could cause the pipe to explode. Venting also helps drainage by allowing air flow. Lack of air would inhibit flow, much like if you put your finger on the end of a straw in a glass of water. Lift out the straw, and the water stays within it as long as you have your finger on the end. Remove your finger and allow air to enter the straw, and the water quickly drains out. Vents also prevent these gases from backing up into the house.

### The Pipes Are Calling

Perhaps the most important decision you'll make about your plumbing—and the one that will most affect your cost—will be the type of pipes you use. Copper may be the first option to come to mind, but you actually have a number of choices available to you:

- **Copper.** Copper is the most popular residential plumbing material, but it's also about the most expensive. It's durable and resists corrosion, but it's not the easiest to install, because it requires soldering to seal the joints (although compression fittings can also be used). Copper piping can be flexible or rigid, but copper piping exposed to freezing temperatures is likely to rupture. In addition, if your water has a low pH level (below 6.5), copper can leach into your water. This can sometimes be corrected with a filter, though.

- **Cast iron.** Cast iron is about the strongest pipe available, but it is also more expensive than many other options. Cast iron's noise-proofing ability makes it the choice of some plumbers, because it reduces the sound of water and waste flowing through the pipes. It is subject to rust, how-ever, which is problematic for many uses.

- **Galvanized.** Galvanized pipe is a metal pipe that has been treated to prevent rust. It was once very popular for supply lines, but it's largely been replaced by copper tubing in most cases. The ends of galvanized and iron pipes are threaded so they screw together.

- **Plastic.** PVC, CPVC, and other plastic pipes are used for residential plumbing systems, for both interior and underground applications. Some concerns exist about plastic piping carrying toxic residue from the plastic into drinking water. Lines that will carry drinking or bathing water should be labeled NSF-PW or NSF-61, which indicates that they meet criteria set by the NSF/ANSI Standard 61 for materials that will come in contact with potable water. Be sure any piping that will come into contact with your water supply has that designation.

  You have different types of plastic pipe available to you. Plastic pipes are relatively easy to install. Depending on the plastic pipe, it may be threaded so it can be twisted together and sealed with joint compound. Waste pipes are flat and can be connected by applying a primer to the ends and sealing the joint with a special cement compound. Waste pipes are generally cleaned with a primer to ensure the seal is as tight as possible.

- **Lead.** Lead pipes were frequently used as water main connections years ago. However, discoveries about the health risks of lead exposure have all but eliminated the use of lead pipes to carry water through homes.

Save money on your plumbing system by using the least-expensive pipe to do the job. Use copper tubing to carry water through the home; use plastic pipe to return waste to the main drain; use galvanized or cast-iron pipe for exterior drain lines and natural gas lines. PVC pipe is also an excellent choice for drain lines because it's durable and won't rust on the inside like galvanized or cast iron might.

## What Size Pipes?

The size of the pipes you use will generally be easy to determine—just check your local building codes. To save money, use the minimum size allowed by code. You can use pipe that's larger than what the building code requires, but your code will give you the minimum standards that must be met. Common pipe sizing is approximately ¾-inch pipe for the main water supply and up to 1½-inch pipe for hot and cold water lines with a reduction to ½ inch near the fixture to create pressure. Drain lines are usually in the neighborhood of 3 inches. Again, check your building code. Typically, the larger your intake pipes, the higher your basic water charges will be, so check with your water supplier to see if this will affect you.

But be careful: don't oversize your pipes to the point that you'll have problems framing around the pipes. For instance, a 4-inch drain pipe can handle more volume than a 3-inch pipe, but a 4-inch pipe will not fit between two 2×4s, which will likely be used to frame your walls. It might be nice to have the additional drainage power, but you don't want to end up with expensive framing corrections or odd bump-outs.

**DON'T TRIP** on your **SHOESTRINGS**

Shutoff valves enable you to stop the flow of water going into your home. The more shutoff valves you have in various places in your home, the easier it is to stop the flow of water to a specific fixture, such as a leaky sink. Insufficient numbers of shutoff valves will increase the number of fixtures and appliances affected if the water needs to be stopped. It's recommended that you have a shutoff valve for each line that runs throughout your home. Plan for shutoff valves at individual fixtures and appliances, each outdoor faucet, as well as to the house and to each floor of the house. Shutoff valves should be easily accessible—you don't want to have to cut through your drywall to turn off the water supply.

## In Hot Water

Your water heater is another essential element of your plumbing system. This, obviously, is going to deliver hot water throughout your home, and you have a number of considerations about how that's done.

### Size

Few things are more irritating than having a head full of shampoo and feeling the shower water run cold. That's why it's important to take into consideration the number of people in your house and the daily demands on the water heater. Rather than looking at the tank's size, look at the heater's first-hour rating (FHR), which tells you how many gallons of hot water the heater can produce during an hour of high usage. Units that use gas and oil systems to fire have higher FHRs than those powered by electricity. The FHR should be close to how much hot water your family uses during peak time—such as morning shower time.

### Location

Your water heater should be centrally located to minimize the time it takes for hot water to reach fixtures and appliances. If you've ever spent several minutes waiting for the shower or faucet water to warm, you know what a big waste of water that is. The closer the hot water heater, the less you have to wait—and waste.

### Type

Water heaters have been evolving over the past few years, and you now have several varieties from which to choose:

- **Storage water heaters.** Usually ranging in size from 20 to 80 gallons, storage water heaters are the most popular but the least energy efficient. When the hot water tap is turned on, hot water is drawn from the top of the tank. Cold water flows into the bottom of the tank, where it is heated, replacing the supply. However, because the water in the tank is constantly being heated, energy can be wasted even when no faucet is on. This is called standby heat loss. Water heaters are getting more energy-efficient, however, so newer models allow less standby heat loss.

- **Demand water heaters.** Also called instantaneous water heaters, these heaters do not have storage tanks. Instead, cold water travels through a pipe in the unit, and the water is heated by a gas burner or electric element when needed. These units eliminate 20 to 30 percent of energy consumption but have limited flow of hot water—up to 2 to 4 gallons per minute. If your family uses hot water at more than one location at the same time (e.g., showering and running the dishwasher at the same time), you may need to install demand water heaters in parallel sequence.

- **Heat pump water heaters.** These water heaters are powered by electricity and move heat from the furnace to the water heater. These systems are expensive to install and should be located in an area that remains between 40° and 90°F. Furnace rooms are a great place to install them, as installing them in a cold space will lower their efficiency.

- **Tankless coil and indirect water heaters.** Indirect and tankless coil heaters use your boiler. The tankless coil system doesn't need an external tank and works well during cold months when the furnace is operating regularly. Tankless coil systems cause your boiler to be operational year-round, though, which uses slightly more energy.

  With an indirect water heater, you do need a separate storage tank. Like the tankless coil, the indirect water heater uses the boiler to heat your water, which then goes into a storage tank. This system is more efficient than the tank-less coil because it doesn't place as many demands on the boiler.

- **Solar water heaters.** Solar water heaters use the sun's energy to heat the water in your home. They come in two varieties: passive and active. Active systems have a pump that circulates the water as well as a temperature control. Passive systems rely solely on the fact that hot water rises and cold water falls and do not use a pump to circulate the water. These systems often have an electric heating component attached that can be used to supplement the hot water supply, if necessary.

### Energy Efficiency

As with any unit, energy efficiency is critical in a water heater. Look at the energy factor number on your water heater to determine the energy-efficiency rating. The higher the number, the more energy-efficient it is to use. Energy-efficient models generally cost a bit more but can return the initial costs in energy savings over the long run.

### Circulation Line

If your water heater can't be centrally located or you want to further increase the speed at which hot water gets to fixtures, you might want to consider putting in a circulating line. Obviously, the farther the fixture is from the water heater, the longer it will take hot water to get to it. The circulation line runs hot water directly to various fixtures and appliances. It has a relatively minimal cost when included with an entire plumbing system, and it can help decrease water waste.

# Planning for Plumbing

Your plumbing will be installed in different steps or stages, and you need to plan for most of them in advance. The following sections list some things you should consider.

## Permit and Paperwork

Obviously, you need to obtain a plumbing permit. Sometimes you can do this yourself, and sometimes your plumber must do it. You also need to apply to your township for permission to tap into the water and sewer mains if you intend to do that. Be sure to allow ample time for your permit and application to be approved. You may also have to pay fees to hook up to public systems.

## Rough-In

The first plumbing stage is the rough-in. This is when you and/or your plumber need to determine the number and types of fixtures and appliances that will need water, as well as the location of supply and drainage systems. (Keep in mind that specialized plumbing fixtures may need to be ordered well in advance.)

Obviously, this is done after the framing but before insulation and drywall are set. If you have extra-large fixtures, put them in place before your walls are framed or you may have trouble getting them through doorways.

During the rough-in, your pipes are run, as well as your main supply and drain. If temperatures are above freezing, turn on the main water line and leave it on, with the shutoff valves closed at each fixture. This will pressure-test your lines and show any weaknesses before you install your drywall. For hot water baseboard heat, use the same technique to pressure-test any lines that have soldered joints and that will be encased inside the walls.

You need to have your rough-in inspected to ensure that everything is up to code and that the system passes inspection. *Do not* put up insulation or drywall before the plumbing inspection is passed.

### Finish

This is when everything is connected and your system becomes operational. Fixtures are hooked to pipes. Your pipes are connected to the main supply and drain systems. The system will be bled of air and should run smoothly without any leaks. It will need to be inspected one more time by your municipality, so be sure to schedule that.

### A Word About Plumbing Inspections

Plumbing usually needs to be inspected at the rough-in stage and at the finish stage. You should plan to be on-hand, if possible, when the inspection takes place so you understand any problems that might crop up.

Generally, inspectors are looking for any potential problems that could cause hazards through leakage or faulty systems. They may examine the size of pipes, number of fixtures, positioning or slope of pipes, fittings, drain heights, and any evidence of leaks. They want to be sure that code requirements were followed in every area of the plumbing installation.

Failing a plumbing inspection means you'll need to correct the problems and have the system reinspected. You should never pay your plumber the balance of his fee until the plumbing inspection is passed.

### It's a Gas

If your home heating system is going to be powered by natural gas or propane, your plumber can also install those necessary pipes. Generally, your gas company will run the line to the

**DON'T TRIP** on your **SHOESTRINGS**

If you buy off-the-shelf plumbing fixtures, be sure the kits are complete. Often, when plumbing fixtures are returned, they are missing parts, which can cause the plumber some headaches and delay—which can possibly result in extra charges for his time. Double-check all fixtures to ensure there are no missing parts.

house (often free of charge if you'll have two or more gas appliances) by digging a trench and laying the pipe, but running the gas to the interior of your home is your responsibility.

We really don't recommend trying to run gas lines unless you're highly skilled at plumbing. Also, your municipality may require that a licensed plumber or qualified HVAC technician do that step. Working with natural gas is extremely dangerous and can cause explosions or subject your family to potentially life-threatening leaks. For the money you'll save, it's just not worth the risk.

Be sure to require the use of galvanized pipes for any exterior gas lines—it's important that your exterior gas line not rust. Interior lines can be copper, cast iron, or so-called "black pipe," or ABS (Acrylonitrile butadiene styrene), a rigid black plastic pipe used for drain lines. There should also be multiple shutoff valves for your gas lines at the various fixtures to which they connect.

### Finding the Right Plumber

Finding the right plumber is similar to finding the right contractor. (We discuss those specifics in Chapter 12.) However, you do need to keep some things in mind when you're looking for a plumber.

Plumbers are licensed only after they complete an apprenticeship, which means that by the time a plumber gets his license, he's got experience. This differs from general contractors' licenses, which have no similar requirement. Therefore, it's a good idea to look for a licensed plumber to do your work.

Also you need to be sure your plumber has both workers' compensation and liability insurance and provides you with proof of both. *Do not* hire a plumber who cannot provide you with proof of insurance. If you do, you are opening yourself to potential liability should someone be injured on your property.

As with any subcontractor, request and check references. Ask these references specific questions about the plumber and his or her work:

- Does your system operate properly?
- Are there leaks or banging noises?
- Does hot water reach fixtures quickly?
- Were there problems with the rough-in or finish inspections?
- How was the plumber to work with?

Speaking with people who've worked with this plumber can tell you a great deal about the quality and reliability of work you can expect.

### Saving Money on Plumbing

There's no doubt about it—plumbing is expensive. In addition to downgrading some of your pipe materials and planning your plumbing so it's centrally located, there are a couple other things you can do to avoid headaches and save money on your plumbing system.

Know your wet walls. Wet walls are walls that have water pipes running through them. It's important that whoever is installing drywall take great care when doing so on a wet wall so they don't accidentally puncture pipes with any screws or nails. Unfortunately, this is an accident that is usually only discovered after the job is finished and water marks start showing up on the wall or ceiling. Use nail plates on pipes in wet walls; these will protect the pipe from being punctured. This is especially important if you've saved money on interior framing by using 2×3s, because the distance between the pipe and the exterior of the wallboard is reduced.

Don't put plumbing in exterior walls without insulation. If possible, keep pipes in interior walls that will be kept warm by interior insulation and the heat of the home.

## Well, Well, Well ...

If you don't have access to a public water supply, you need to dig a well on your property to get water to your home. How difficult, how deep, and how expensive this process will be largely depends on your location. Speak with the people at your municipal offices to get some information on what you might be able to expect. Your building inspector can likely give you the rundown on how deep other people in your area have had to dig to reach a suitable water supply.

When you hire a contractor to dig a well, you're usually going to be charged by the foot. However, because many contractors won't guarantee hitting water, you essentially pay for however deep the well needs to be. If the contractor hits water, it needs to be tested to ensure that it's suitable for drinking and bathing. If not, the well may need to be dug deeper. Discuss a depth limit with your contractor up front. If he or she reaches that limit and hasn't hit water, you can discuss the next level. However, you don't want to end up being surprised by a digging bill that is twice what you expected.

A few other considerations when digging a well include the following:

- **Water treatment.** After your water is tested, you may need to take corrective action. Depending on the quality and mineral content of the water (water with high levels of dissolved minerals is called "hard water"), you may need a filtration or water treatment system to improve the water's quality.

- **Location.** Your well should be within 50 feet of your home, if possible. However, it should also be located away from septic systems to avoid accidental contamination. If your septic system is on one side of your home, try to put your well on another side, with the well placed on higher ground than the septic system, if possible.

- **Frost line.** If you live in a cold climate, be sure the main that runs water from the well to your home is placed below the frost line.

- **Roots.** Be sure no large trees are located near the well, because roots can collapse the interior of the well.

- **Pumps.** Well pump strengths and features vary depending on the depth of your well, the amount of water you need, and other factors. Discuss with your contractor the best pump for your needs.

## Septic Matters

If you don't have access to public sewers, you'll likely need to install a septic system to dispose of plumbing waste. Your local building code has strict rules about installing your septic system, and you need to follow those rules.

A septic system connects to the main drain in your home, and waste from that drain is deposited into a septic tank. Your tank can be coated steel or concrete (the latter is more expensive but more durable). Enzymes and bacteria in the system break down the material in the tank, and the remaining liquid is then distributed through an outlet into a leach field. The liquids are then distributed, through perforated pipes, into the ground, where they are absorbed and filtered by layers of soil.

Because septic material is released back into the ground, it is essential that you place your septic system away and downhill from wells that supply drinking water. Also be sure that you choose the appropriate-size tank for your home. Too-small tanks could allow sewage to back up into your home.

The size of your septic tank will be determined by the number of bedrooms in your home. The exact size will be in your local building code, but you can expect a tank of approximately 900 to 1,200 gallons for a 3- to 4-bedroom home. Don't skimp here—it's essential that you have a properly sized tank for your home. In general, we recommend having a qualified plumber or septic technician install your system, because improper installation could be hazardous to your family and those in your community.

## It's Electric

Your electrical system is another critically important system in your home—and in demand more today than ever before. Multiple computers, media stations, appliances, and myriad other electricity-requiring devices make planning this system more important than ever before.

You may have an electrical plan that came from your architect or with your stock plans. If you don't, your electrician can help you develop one.

This is another area where doing it yourself is dicey, unless you really know what you're doing. Your municipality may also have requirements about who runs the electricity to your home. If you really feel as though you want to try it on your own, consider running the wiring yourself (if you're sure your plan is appropriate and up to code) and having a licensed electrician hook up the wiring to the service box.

In your planning stage, as we discussed, pay attention to how each room will be used. If you plan on having rooms that will utilize a high volume of electricity, such as a media room or a

**DON'T TRIP**
on your **SHOESTRINGS**

Do not plant trees directly over or very near to your septic system, as the roots can potentially damage your system and cause expensive repairs.

home office with multiple computers, you should share that with your electrician. He or she can be sure these rooms are properly wired.

Building codes play a key role in your electrical plan. You'll likely be required to have an outlet every specific number of feet along the wall, use specific types of wire in your system, and install ground fault interrupter (GFI) outlets in kitchens and bathrooms near water supplies. (These outlets break the flow of electricity immediately in cases where there is a surge.)

Even if you don't intend to do it now, tell your electrician if you plan on installing a swimming pool, spa, or other outdoor feature that will require electricity. He or she can make accommodations for those future expansions in your electrical system.

### Amp-ing Up

The amount of electricity available for use depends on the size of the service box, or breaker box, you use. The service box is the rectangular board of switches that controls the flow of electricity to various areas of your home.

The average home needs a 240-volt, 100-ampere service, although 200- or even 400-ampere boxes may be required in larger homes or those that have a higher number of electrical fixtures, including a swimming pool, multiple electrical appliances, and the like.

Within the service box are circuits, where wires connect to the service from throughout the house. Another reason you should discuss the various electrical needs in each room is so your electrician can properly plan various circuits so that no one circuit has too many devices drawing on it. Lack of circuits is the most common reason electrical systems need to be upgraded. If you run out of circuits, you may have to add an additional circuit box later, at additional expense. The service box needs to be grounded (this is a safety mechanism to ensure you don't get shocked when you're checking for a tripped circuit).

"No new wires" is a mantra of the National Electrical Contractors Association, and we agree. It's better to have too many wires snaking through the walls of your home than too few, because rewiring your home is an expensive proposition. It's much cheaper to run wires while the walls are open and clear of obstacles.

### Parts of the Electrical System

In addition to the service box, the electrical system is made up of several parts:

- **Wiring.** All wires are not created equal, and you'll likely need several different types of wiring for your home. Heavier-gauge wires are used for appliances, whereas lighter wires can be used for circuits for lighting fixtures. Check your local building code for specific requirements.

- **Outlets.** Duplex outlets are those standard, two-plug outlets that can be found along walls in any home with electricity. Again, your local building codes will have requirements about how many outlets need to be installed in your home and at what intervals. GFI outlets, which act as mini-circuit breakers, may be required around water sources and outdoor outlets.

- **Switches.** Switches control the flow of electricity to lights and other devices. Switches can be standard "on/off" or dimmers, which allow you to control the level of light. Typical switches are "one-way" switches, meaning they allow one point from which to turn the flow on and off. With two- and three-way switches, you can turn the device on or off from multiple areas.

- **Lighting.** Plan your overhead and exterior lighting before you start wiring. You must decide whether your lighting will be recessed, which is a more expensive option than traditional fixtures but can conserve space if you have lower ceilings. Be sure recessed lighting is insulated properly to avoid drafts and a potential fire hazard when lighting could overheat insulation that comes in contact with it.

- **External power sources.** You'll probably need a few outlets outside, especially if you plan on having a swimming pool, decorative lighting, or other outside devices requiring electricity. Discuss these requirements with your electrician.

### Low-Voltage Wiring

Your electrician should also have knowledge of low-voltage wiring that will carry sound, data, and video, such as the following:

- Telephone
- Cable television
- Burglar alarm
- Video surveillance
- Intercom
- DSL or cable Internet connection
- Doorbell
- Home automation system

In an increasing number of municipalities, smoke detectors must be wired into your electrical system, so be sure to plan for them if your area has such a requirement.

It won't add much to your cost to run these wires now, but if you try to do it after the walls are sealed with drywall, you will incur costs. In addition, when it comes to resale value, this wiring will become critically important. With the increasing demand for home automation, failure to install simple CAT5 wire (the wire used for many voice and data connections) and standard cable can actually make your home obsolete.

## Electricity in Action

Just as you needed to apply for permission to connect to public water and sewer systems, you need to apply—and pay for—permission to tap into the public power system. Next comes the rough-in of your electrical system. Similar to your plumbing rough-in, this is where your wiring is run and outlets and switches are installed. You need to have this rough-in inspected, so *do not* begin drywall installation before your electrical rough-in has passed inspection.

Later, after the drywall and the flooring are in, comes the finish phase, when all your big appliances are connected and all the wiring is completed.

## Finding the Right Electrician

The guidelines for finding an electrician are very similar to the guidelines for finding a plumber, discussed earlier in this chapter. You can also use the good business practices we discuss in Chapter 12. Look for a license, because most electricians need to complete an apprenticeship, much like plumbers do. Check references and request bids from at least three electricians to have a basis for cost comparison.

## Saving on Your Electrical System

The best way to save on your electrical system is to plan in advance. It's likely that your electrician is going to charge you based on the number of outlets and fixtures you have. You can save by leveraging a competitive bid and asking for a better price.

We are firm believers in leaving much system work to professionals, especially something as potentially dangerous as hooking up your electrical system, but running wires isn't a terribly difficult thing to do. If your electrician is game, see if he or she will let you drill the wire holes and even run the actual wires throughout your home. You need to be sure you know which wires go where, because it will end up costing you more money and time to go back and correct any incorrectly run wiring. But this is an area of potential savings—how much depends on the contractor and what you're able to negotiate.

You may want to purchase wire, fixtures, plates, and the like yourself to save the electrician's markup, but check his or her prices as well—the professional discount your electrician gets may be cheaper, even with markup. If you do purchase them yourself, be sure you have the proper specifications and that the materials arrive on time. Otherwise, you could end up delaying your project.

## Running Hot and Cold

After plumbing and electricity, the third system critical to your home is your HVAC system. You probably have more options here than in any other system in your home.

### Power Source

First, you need to decide the power source you'll use to heat your home. You have several choices for fueling primary heating systems, and most people have a strong preference for one source over another:

- **Natural gas** Not available in all areas, natural gas is delivered through a main. It is considered a "clean" fuel, because it doesn't produce residue when it burns. Rates are set by the utility company, but overall, this is an affordable heating source.

- **Fuel oil.** Fuel oil is easy to access. Prices depend on the marketplace, but again, it's relatively affordable. Critics of fuel oil say it isn't as clean a resource as natural gas, but improvements in furnace systems have limited the odor and residue that was once associated with fuel oil. Oil is stored on your property in a tank. It's critical that the tank be inspected and meet strict safety standards. You should also obtain insurance against leaks, because fuel oil leaching into the ground is a very expensive problem to clean up—to the tune of tens of thousands of dollars.

- **Electricity.** Electricity is a readily available power source and may be a good option for warmer climates or vacation homes that only need heat on a limited basis. Using electric-powered heat eliminates the need for a furnace or boiler, but heating a house solely by electricity in a cold climate can be very expensive, so it's not the best option for those areas.

- **Propane.** Propane, or liquid gas, has the same benefits of natural gas, but it comes in a liquid form that is stored in a tank on your property.

- **Solar.** Solar fuel systems convert sunlight into electricity. The challenge comes when there isn't enough sunlight to create a sufficient energy supply, so solar systems need backup systems, which are usually electric. The good news about solar fuel systems is that you may be eligible for significant tax credits or government rebates. Even though the system could cost as much as $40,000 or more to install, making it the most expensive fuel source, you'll likely save at least that much.

Wood and coal are also considered heating sources, but we don't recommend using them as your primary fuel source.

## Heating Up

When you've decided on a fuel source, you'll want to consider your heat-delivery system. Water-based systems are powered by a boiler, which heats the water running through the system, whereas a furnace is used for hot air–based systems. Again, you have several choices:

- **Forced air.** In forced-air systems, hot air is carried from the furnace through the duct system, blowing warm air through supply vents in each room. In each room are registers, or returns, usually near doors or windows, which bring cooler air back through ducts to the furnace, where the air is reheated.

  Forced hot air is an economical choice for heating and cooling systems because the same ductwork and air handlers can be used for both, eliminating the need for separate heating and cooling systems.

- **Hot water baseboard.** With a hot water baseboard system, a boiler heats water, and the water is then carried through pipes (usually copper) housed in baseboard units. Heat from the hot water radiates from the units, heating the room.

- **Electric baseboard heat.** In this system, baseboard elements within the unit warm and give off heat. Using electric baseboard units requires more electrical service to your home and is not generally a good option for cold climates because it can be expensive. For warmer climates and homes that don't need frequent heat, this type of system eliminates the expense of a boiler or furnace.

- **Radiant.** Radiant heat is similar to hot water baseboard systems, but the hot water is heated while it's circulated through pipes in the floor. Because warm air rises, the heat moves upward and cool air falls, where it is warmed by the heat coming off the floor.

- **Heat pumps.** Heat pumps come in two types: air-to-air and geothermal. Geothermal, or ground source heat pumps (GSHPs), are electric-powered systems that force air from the home into the earth, where it is warmed and then delivered back to the home. Air-to-air heat pumps force air into an exterior unit, where it is heated and then delivered back into the home.

Whichever system you choose, your house will be warm when it needs to be.

## MONEY IN YOUR POCKET

If your budget is running low, consider using forced air heating, which utilizes the same ductwork as your air conditioning system, saving you the expense of installing additional heat conductors to each room. That could save you anywhere from $2,000 (for a small house) to $20,000 or more (for a large home), depending on the heating system you choose. (We used the low-end figure of $2,000 for the running total.)

**Savings for You: $2,000 to $10,000  Running Total: $55,430**

### Cooling Down

Although it's feasible that you can cool your home with fans or window air-conditioning units, central air conditioning has become relatively standard in most areas and will add to the resale value of your home.

A central air-conditioning unit works much like a heat pump. Air is forced from your home into an external unit called a compressor. The compressor cools the air, and the cool air is then forced back through ductwork in your home to cool each room.

If you couple your air conditioning with a forced hot-air heating system, you'll maximize your heating and cooling investment.

### Venting
Your home needs a ventilation system. Proper ventilation allows fresh air to enter the home, reducing interior pollutants. Fans located throughout the house, in the attic, and especially in moist areas such as bathrooms vent air out of your house.

### Zoning Out
Whatever heating and cooling systems you choose for your home, they will be connected to and controlled by thermostats, which dictate the temperature of the home. You can save on your energy bills by creating zones for your heating and cooling and closely controlling those temperatures. For instance, if your downstairs family room and kitchen are zoned differently from your upstairs bedrooms, you can heat or cool the area of the house the family is using and adjust the temperature in the empty portion of the house to a more energy-conscious level.

Zones are less expensive to install for hot-water baseboard heat, because the same boiler can be utilized by adding a pump. Federal Housing Administration (FHA) requirements call for a separate air handler for hot air–based systems, however, so failure to install a handler could affect your qualification for FHA financing when you refinance your construction loan.

### Saving Money on Heating and Cooling Systems
Heating and cooling systems will take up a large portion of your budget. However, you can save some money:

- **Work with your local utility companies.** Your local utility companies will conduct an energy audit and give you tips to save money on your systems design, as well as ideas to increase energy efficiency, which will save you money over the long haul. They may also offer rebates for new systems that meet certain efficiency requirements, so call your local gas, electric, or other energy provider to find out what programs are available to you.

- **Shop at auctions.** Public auctions are often held when companies go out of business or to auction off the property of individuals for various reasons. A friend of ours went to one such auction and got thousands of dollars worth of plumbing fixtures for a few hundred dollars. Look for these auctions in the "Public Notice" section of your newspaper.

- **Buy equipment off-season.** Buy your equipment when it's not in demand, and you'll often get a better price. Air-conditioning compressors are often less expensive in December than they are in June.

- **Scour scratch-and-dent sales.** Slightly damaged equipment can save you big bucks on appliances, fixtures, and devices.

## Dollar-Saving Do's and Don'ts

- Check and double-check local building codes, which will have strict guidelines for installation of all your home's necessary systems.

- Mix the types of pipes you use to save money, but maintain the best function.

- Design your systems with energy efficiency in mind.

- Whenever possible, keep systems centrally located to minimize the lengths heat and hot water have to travel. This will also minimize the lines that need to be run.
- Take advantage of auctions, sales, and off-season opportunities to get the best prices on materials and system components.

# 9

# INSIDE JOBS

As your home comes together on the outside, you'll have many decisions to make and jobs to complete on the inside of the structure.

## Plan for Usage

One of the key areas where you can save money on your home is to plan its usage and make decisions accordingly. Take some time to think about each room and the activities that will be going on in that room. Will your family room be a media center with many electronic pieces, cable, and a computer with Internet access? If so, it will probably need extra wiring. The same goes for media rooms and home offices or any other rooms that will have special elements such as central vacuums, intercoms, etc. It's cheaper to make these accommodations when the home is being built than it is to hire a contractor to make the additions later.

Similarly, put durable, easy-to-clean floor and wall coverings in areas that will be heavily trafficked, such as hallways, the family room, or the kitchen. This will save you money on replacing less-durable materials in the future.

# Building In

In addition to the external framing on your house, you need to consider interior framing and building. The more interior detail you have, the higher your framing costs are likely to be. However, some of these built-ins do give a return on your investment and you can save money on them:

- **Interior walls.** Your interior walls are framed similarly to your exterior walls. Look for walls that create "dead" space and unnecessary walls. You may find that a wall has been placed in an area to create a specific shape. For instance, in Hank and Gwen's home, a few small walls closed off corner areas. Because these were not support walls and were only put in to make rooms and foyers a particular shape, we eliminated them, creating interesting nooks and crannies that could be decorated with small tables or plants. Eliminating these walls saved some money on framing, materials, and finishing.

- **Closets.** Closet space is a big factor in resale value and will make your home easier to keep clutter-free. As you frame closets, keep in mind the type of doors they will need. Bi-fold doors are generally more expensive than traditional doors, but they can offer greater access to a longer width, so the convenience may be worth it.

- **Room dividers.** Instead of using a full wall to separate rooms, try incorporating half-walls, columns, or built-in shelving to separate rooms. This creates a more-open floor plan and will save you money on labor, materials, and finishing.

- **Shelving.** Built-in shelving can be lovely and offer great storage. Instead of having custom shelves built, you could save money by choosing stock or semi-custom shelving options from your cabinet provider. Depending on your needs and the space in the room, you could end up saving hundreds or even thousands over customized shelving, which can be very expensive.

## Upstairs, Downstairs

If you are building a home that has two or more stories, you need to think about stairs. As with most things, the simpler you make the stairways in your home, the cheaper they will be to build.

Because stairs must be well-built to avoid creaking and also must be precisely sized to be consistent in height—which will make them safer—we strongly recommend hiring a professional to build your stairs, unless you are a skilled carpenter.

Stairs consist of *treads,* which are the horizontal areas that you step on, and *risers,* which are the vertical areas that boost the next step up. Your stair treads should always be constructed of hardwood such as oak, but to save some money, your risers can be constructed out of a softer wood such as poplar, especially if they will be painted. If you're going to carpet your stairs, you can use pine, which will be protected by the carpeting.

If your stairway is curved or irregular, it's going to cost more than a straight run of stairs. Spiral staircases take up less room but are usually the most expensive type of staircase and will decrease some of the function of your stairs. (Just try moving a sofa up a spiral staircase!)

### Rise and Run

It's best to have the carpenter who will be building the stairs come out and measure the area of the staircase exactly, but some of the measurements you need to consider when building stairs include the following:

- **Total rise.** This is the height, vertically, from the floor where the stairs begin to the floor where they end.
- **Total run.** This is the horizontal measurement from where the stairs begin to where they end, and measures the length of horizontal distance the stairs cover.

The ratio of the total rise to the total run gives you the slope of the stairway. Usually, stairways should be about 30 to 35 degrees in slope, but check your local building code requirements. It's important that stair rises—the height of each stair—be uniform. Inconsistent rises can be dangerous, as users will be more likely to trip. You should also be sure that the stairs meet the height of the flooring at the top of the stairs evenly, or that could create another safety hazard.

## Width
Ideally, stairways should be a minimum of 36 inches wide, preferably, if you have the room. That's wide enough for two people to pass and also makes hauling furniture and bedding upstairs easier. (More than one king-size bed has had to be scrapped because the mattress wouldn't fit up the stairs!)

## Railing
You also need to be sure that your stairway has a proper railing, which is properly anchored to the wall or connected to the stairway by a series of balusters, or spindles. Generally, these railings should be about 36 inches high, but they can usually be adjusted if members of your family need them a bit higher. Check your local building codes, which are often very specific about railing requirements.

Railing materials range from wrought iron to wood spindles and virtually every material in between. They should be sturdy, and, if you have small children, the spindles should be close enough together that curious little heads don't get caught between them.

## MONEY IN YOUR POCKET

Check out the savings you can enjoy from buying manufactured staircases rather than having them built on-site. Depending on the staircase, you could save 10 percent or more. On a $2,000 staircase, which is a moderately priced staircase, that's $200.

**Savings for You: $200          Running Total: $55,630**

### Attic Stairs

Don't forget attic stairs, which may be either a full staircase, or more likely, less-expensive premanufactured pull-down stairs. Pull-down stairs, or folding stairs, fold up into the ceiling and open like an accordion. They're generally narrow—approximately as narrow as 22 inches wide, so they're not a great option if you're going to be hauling lots of stuff in and out of your attic.

Which staircase you choose will depend on the access area to your attic, as well as whether your attic has much functional space. Hank and Gwen put a 36-inch stairway to the attic of a home they rebuilt. This allowed for easy access to put furniture, boxes, and the like up there, plus gave us the option to turn the finished attic into a study or other living space.

Attics formed from wood trusses, on the other hand, have limited storage space. And if the lower beam used to make the wood truss is a 2×4, it's likely that the attic floor won't be able to support much weight anyway, so a pull-down stair will be a less-expensive option—a few hundred dollars vs. as much as $2,000 to $3,000 for a simple traditional stairway. Just be sure to create a space for the stairway or that it fits between your ceiling joists to avoid costly framing modifications.

### Basement Stairs

If you have a basement, you also need stairs leading down into it. Basement stairs can be built from wood or concrete, or you can use steel or premanufactured stairs. Your stairs may lead from the interior of your home to the basement floor or, if you have a basement only accessible by exterior doors, they'll lead from the doors to the basement floor. (In this case, the stairs are usually concrete because they are often exposed to more of the elements.)

As you plan your stairs, be sure to consider the basement's use, both now and in the future. If you have plans to finish the basement and make it into living space, you will want to build your stairs so they are attractive enough and functional enough to fit in with that usage.

## Insulation 101

When your home is closed on the exterior (framed, sheathed, and sided), it's time to think about insulation. Proper insulation provides a number of benefits, including increased energy efficiency and sound-proofing.

According to the U.S. Department of Energy (DOE), the typical U.S. household spends 44 percent of its utility bills—close to $1,300 each year—for heating and cooling costs. The DOE estimates that homeowners can reduce that cost between 10 and 50 percent per year, depending on the home and its construction, the energy efficiency of heating and cooling systems, the area climate, and the family's energy-conservation habits.

### Understanding R-Value

The first thing you need to know about insulation, however, is its R-value. According to the Insulation Contractors Association of America (ICAA), R-value measures insulation's resistance to heat flow and is also called "thermal resistance." The higher an insulation's R-value, the greater the insulating power. When buying insulation, be sure not to get sidetracked by the thickness of the material. Whatever the type, thickness, or weight of an insulation product, all products that have the same R-value offer the same insulation value.

### Types of Insulation

According to the ICAA, there are a number of different types of insulation:

- **Fiberglass.** Probably the most popular form of residential insulation, fiberglass is made from molten sand or recycled glass and other inorganic materials under highly controlled conditions. It can be used in interior and exterior walls, as well as ceilings.

  Fiberglass is produced in batt, blanket, and loose-fill forms (one popular brand is pink). Batts and blankets come in rolls, often with a kraft paper or foil moisture barrier (the former is less expensive). You can also buy it without a moisture barrier (cheaper still) and use a sheet of plastic as the barrier. The loose fill can be spread or blown into your attic using an insulation blower.

- **Cellulose.** Cellulose is a loose-fill insulation made from paper to which flame retardants are added. It's usually blown into walls or shaken directly from the bag into open areas, such as in the attic. You may incur an extra cost with cellulose if you have to rent an insulation blower. Also cellulose is more susceptible to moisture damage and, when it packs down, can lose some of its R-value.

- **Foam insulation.** Foam insulation comes in two forms: a rigid board or a liquid foam that can be used to fill in and seal small areas and cavities. The rigid board insulation is usually used on the exterior of the home. Foam insulations are not fire resistant.

- **Rock and slag wool.** Rock and slag wool are manufactured similarly to fiberglass but use natural rock and blast furnace slag as its raw material. Typical forms are loose-fill, blanket, or board types.

- **Reflective material.** Reflective materials are fabricated from aluminum foils with a variety of backings such as polyethylene bubbles and plastic film. Reflective insulations retard the transfer of heat. They can be tested by the same methods as mass insulation and, therefore, assigned an R-value.

### How Much Is Enough?

The R-value you use for your home will vary according to a number of factors, such as your climate, the type of heating and cooling you use, and the energy efficiency of your home's design.

The DOE has geography-based guidelines on recommended R-values on its website (www.eere.energy.gov/consumerinfo/ energy_savers/r-value_map.html). Obviously, the colder the climate, the higher the R-value. You should also be sure your insulation plan meets your local building code. The DOE also includes local codes on its website (www.energycodes.gov/ implement/state_codes/index.stm).

It doesn't pay to go much beyond the recommended R-value in your area. In moderate areas, such as the Mid-Atlantic region, R-38 is recommended in external walls. However, in cold areas such as the northern Midwest, recommendations leap to R-60.

Insulation that's too thick for the space will actually end up losing R-value because compression will force out the air within the insulation, which helps regulate temperature. Do opt for max R-value in interior walls and floors, however, because it will create terrific, low-cost sound-proofing.

### Insulation Installation

Installing standard bat or blanket insulation isn't a particularly difficult task, in most instances. Doing it yourself can save you hundreds of dollars in installation fees. However, it's not fun. It can be messy, and many insulation materials, especially fiberglass, can irritate the skin, eyes, and lungs. Be sure to wear a particle mask and gloves. It's a good idea to wear long sleeves and long pants, as well. Goggles are also recommended.

If you go the contractor route, the ICAA recommends that you be careful of any contract or verbal offering that quotes the job in terms of thickness only (e.g., "10 inches of insulation"). It's the R-value—not the thickness—that tells how well a material insulates. When you have chosen an insulation contractor, be sure the contract includes various job specifications, cost, method of payment, and warranty information provided by the insulation material manufacturer. Be sure the contract lists the type of insulation to be used and where it will be used. Be sure each type of insulation is listed by R-value. Avoid contracts with vague language such as R-values with the terms "plus or minus," "+ or –," "average," or "nominal."

## On the Wall

After your insulation is in place, it's time for the walls to go up. As with nearly every other part of the homebuilding process, you have some choices in the type of walls.

## Drywall

By far, the most common form of wall material is wallboard, often known as drywall. This material comes in large boards, about 4×8 or 4×12 feet, usually made of gypsum plaster housed between two sheets of paper. The boards are nailed or screwed to the wall studs. Then the seams are sealed using a special tape and covered with a compound that is allowed to dry. The combination of the tape and compound are then sanded. The process is repeated two or three times, and the desired finished effect is a perfectly smooth, consistent surface.

Drywall generally comes in three types:

- **Standard.** This is your garden-variety drywall that's used in most areas in the home.
- **Water resistant.** Used in bathrooms and other moisture-rich areas, it's sometimes called "green board" because the exterior is usually green or blue in color.
- **Fire resistant or type X.** This kind of drywall is sturdier and, as the name implies, doesn't burn easily. It's usually required on the walls where an attached garage meets the house.

## Saving Money on Drywall

The more rectangular the room, the lower your drywall cost is going to be, both in terms of labor and materials. Wallboard starts getting expensive when you have vaulted ceilings, curves, wide expanses (such as open foyers or family rooms), many cuts and angles in your room, or exceptionally high ceilings. The more complex your design, the higher your drywall cost will be, so the best way to save money is to keep it simple.

Aside from simple room design, the easiest way to save money on drywall is to avoid waste. Discuss with your contractor how he will install the drywall and which size board is best

**DON'T TRIP** on your **SHOESTRINGS**

Consider using nail guards to protect your plumbing and electrical systems. These small plates ensure that stray nails or drywall screws don't puncture any system components, requiring expensive repairs. The small investment they require may save you headaches and expense later on.

for his plan. Why buy 4×8 boards if 4×12 boards will fit the room better, creating fewer seams and less labor to fill those seams?

### Paneling

Paneling is another type of wall material that can add warmth and impact to a room. Wood paneling generally comes in a variety of widths and finishes, usually with a tongue-in-groove fit. Paneling can be applied over drywall or directly to the studs.

### Plaster

Although it's not often used in new construction, we do need to discuss plaster briefly. Plaster is a mud or pastelike material that is applied over boards of lathing (strips of wood) that helps hold it in place. The plaster is applied, smoothed, and allowed to dry. It's much more expensive than drywall and is more prone to cracking, but if you're creating a very traditional home, it's a nice touch.

## The Great Wall Cover-Up

Now that you have gorgeous, pristine walls, you'll notice that they're so ... white! It's time to let your inner interior designer out by choosing the wall coverings that will give your dream home warmth, character, and drama.

As you decide the best coverings for your walls, keep this in mind: it's best to first prime the walls with a good-quality primer. This will protect your drywall by sealing it and will also make wallpaper adhesive easier to remove. You often need less paint to fully cover the wall when you apply a primer coat first.

### Paint

Paint is the least-expensive material to use to cover your walls. Be a savvy paint shopper, and buy better paint on sale instead of buying cheap paint, which often won't wash well.

Go to your local home store, and you'll find a great number of brands and types of paint. For general information, the National Paint and Coatings Association (NPCA) is a good place to start. The association lists the various brands and types as follows:

- **Latex paints.** These are water-thinned and apply easily with a brush or roller. Cleanup with soap and water is a distinct advantage. Latex paints are available in most gloss ranges and will do a good job in most interior areas. They are not flammable and have a very mild odor.

- **Alkyd (oil) paints.** These are solvent-thinned paints. They apply well with a brush or roller but need turpentine or mineral spirits for cleanup. Alkyd paints are sometimes preferred for areas where constant cleaning is necessary, such as kitchens and bathroom shower areas. Very-high-gloss enamels are usually solvent-thinned. The odor is stronger during application than with latex paints, but disappears after a few days.

- **Enamels.** Enamels are generally smoother and dry to a harder surface than other interior paints. They are available in high or low gloss and can be either latex or alkyd.

- **Gloss.** The gloss is the luster or shininess of a dry paint. Paints are usually classified as flat, eggshell, semi-gloss, or high gloss. A wide variety of gloss ranges is available.

- **Special paints and coatings.** These are available for most surfaces. Wood floors, concrete, or masonry and metal surfaces require specific products. Consult your paint retailer and read the paint can label carefully for recommendations.

## Saving Money on Painting

The NPCA says that, although you can use almost any paint for drywall, masonry usually contains alkali, so be sure any paint you use on masonry work is akali-resistant. Also, over iron or steel, opt for a rust-inhibitive primer. You can use any type of enamel or paint over the primer as a topcoat, depending on the use of the area to be painted.

Latex paints are usually the most affordable. If you plan on using rich, dark colors, you can save money by tinting your primer instead of using white primer. Also, if you see a color you love in one of those expensive designer paints, you can always take the chip to a different paint vendor and have the color matched in a less-expensive product.

If you are someone who likes to change paint colors frequently, you can save money by choosing a lower-quality paint, because it won't need to last for a number of years. As always, shop around. Paint is always on sale somewhere, and often you can save as much as 40 percent, depending on the retailer. Opt for the largest size of paint container you need—gallons over quarts, 5-gallon buckets over gallon containers. The more you buy, the more you save.

## Faux Finishing

A great way to get some of the dramatic effects of wallpaper without the expense is to use faux finishing. A variety of faux finish techniques are available, and most are achieved by combining various colors of paints with different applicators, such as cheesecloth, special rollers, sponges, and the like. Although the increased cost of materials and applicators make these projects more expensive than painting, they are far less expensive than wallpaper.

## Using Wallpaper

Wallpaper can be a gorgeous way to cover your walls. From very subtle patterns to dramatic murals, wallpaper delivers drama and durability. It can cover flaws in your walls and has a rich look.

Like most home building and décor materials, there are many different types of wallpaper. According to the Wallcoverings Association, the most popular types of wallpaper you will find at your local wallpaper retailer or home improvement store are as follows:

- **Vinyl-coated paper.** This wallpaper has a paper substrate on which the decorative surface has been sprayed or coated with an acrylic-type vinyl or polyvinyl chloride (PVC). These wallpapers are classified as scrubbable and strippable and are suitable in almost any area. These papers are more resistant to grease and moisture than plain paper and are good for bathrooms and kitchens.

- **Coated fabric.** This type of wallpaper has a fabric core that is covered with liquid vinyl or acrylic. The decorative layer is printed on this coating. Coated fabric wallpaper is more "breathable," which makes it best for use in low-moisture rooms such as living areas.

- **Paper-backed vinyl/solid-sheet vinyl.** This type of wallpaper has a paper (pulp) substrate, or core, which is laminated to a solid decorative surface. This wallpaper is very durable, because the decorative surface is a solid sheet of vinyl. It is classified as scrubbable and peelable. Solid-sheet vinyl can be used in most areas of the home, because it resists moisture and is stain and grease resistant. However, this type of wallpaper will not withstand hard physical abuse (such as what it would get in mudrooms or storage areas).

- **Fabric-backed vinyl.** Fabric-backed vinyl has a substrate laminated to a solid vinyl decorative surface. The most common kind of product in this category is solid vinyl, which has a vinyl film laminated to a fabric or paper substrate. Vinyl is more durable, but paper is more affordable.

Buy your wallpaper from the same dye lot or run number so the colors are consistent. Before you buy (or hang) anything, check each roll to ensure uniformity of tone, colors, and pattern.

## Wainscoting

Wainscoting is a wood wall covering that extends about halfway up the wall. It comes in a variety of styles and can be painted or stained, depending on your preference. It adds a rich feel to the room and can protect walls from damage.

Newer laminate wainscoting comes prefinished and can be applied with nails, glue, or both. This wainscoting is less expensive than wood wainscoting and doesn't need to be painted or stained. You will need to fix nail holes if you use nails, though.

For a less-expensive option, do faux wainscoting—use a different color paint or wallpaper on the bottom 3 feet of the wall. Then, set a chair rail molding around the top of the area to finish it. It's a great look that's similar to wainscoting, but much less expensive.

## Molding the Space

Molding, also called trim, is generally installed at the lower joints of the room, where the wall meets the flooring. It covers the transition from wall to flooring and can also be used to do the same at the ceiling, where a more decorative type of molding, called crown molding, is usually used. Crown molding is also used around windows and doors to hide drywall edges.

It can be used to add decorative touches on walls and other areas, as well, such as faux picture frames on walls, chair rails, and the like.

Molding comes in a number of styles, sizes, and materials, and choosing the right type depends on the purpose for which it will be used. If you're planning on staining your molding rather than painting it, opt for top-of-the-line oak or clear pine wood molding. If you're planning on painting the molding, you can choose less-expensive molding made from composites or laminate materials, which will be virtually indistinguishable from wood after the paint is applied. Regardless of material, the more decorative the molding, the more expensive it will be.

Molding is one of the easiest things to put off to cut cost. If you're watching your budget, choose less-expensive molding, and save crown moldings, chair rails, and other touches for later. They're easy to add later, and putting this off could save you up to a thousand dollars or more, depending on the size of your home and the type of molding you plan to use.

## MONEY IN YOUR POCKET

Consider using drywall casing as the finishing around windows and doors. The drywall will sit flush against the window jam and is finished so it doesn't need any molding around it. Fully cased windows—those with molding surrounding the window—are more expensive in both materials and labor. By saving $25 in materials and labor in a home with 16 windows, you've pocketed $400.

**Savings for You: $400**          **Running Total: $56,030**

## Floor Coverings

Your floor coverings will be one of the first interior design decisions you make about your house. When you're considering which floor covering makes sense, consider the use of the room and your particular circumstances. Highly trafficked areas should be covered with durable, easy-to-clean materials. Save the more expensive and delicate flooring for rooms that are used less frequently. For instance, it may make sense to carpet the family room in a durable carpet that won't show dirt and stains, whereas a light-colored, more expensive carpet might make your parlor or living room a showplace.

The majority of flooring falls into five categories: vinyl, hardwood, laminate, ceramic or stone, and carpet.

### Vinyl

Vinyl flooring has come a long way since your grandmother's day, and today you'll find many attractive options from which to choose. Vinyl flooring comes in sheets and tiles—tiles are easier to install, but sheets help you eliminate some, if not all, seams, as they cover larger areas. Vinyl is usually the least-expensive flooring option.

If you can, opt for parameter bond instead of full-spread adhesive flooring. Full-spread adhesive means that the flooring has a layer of adhesive underneath it.

### Hardwood

Wood floors are crafted from hard species such as oak, poplar, hickory, cherry, and the like. Softer woods, such as pine, are also used, but we don't recommend them because they are more easily damaged.

The National Wood Flooring Association (NWFA) describes the various styles of wood flooring as follows:

- **Solid flooring.** All wood flooring, regardless of width or length, that is one piece of wood from top to bottom is considered solid flooring. Solid flooring gives you a great opportunity for customization. Your choice of type of wood, stains, and finishes all contribute to the personalization of a solid floor. This is an excellent choice in most areas of a home on the ground level or above.

- **Engineered flooring.** This wood flooring product consists of layers of wood pressed together, with the grains running in different directions. It is available in 3 and 5 ply. Engineered flooring is perfect for those areas of the house where solid wood flooring may not be suitable, such as basements, kitchens, powder rooms, and utility rooms. Because the grains run in different directions, it is more dimensionally stable than solid wood.

- **Acrylic-impregnated floors.** Acrylic-impregnated floors are created using a process in which acrylics are injected into the wood itself, creating a superhard, extremely durable floor. This type of flooring is often used in commercial installations such as shopping malls and restaurants; however, they are right at home in busy households as well.

Wood floor types come in three basic styles. *Strip flooring* is linear flooring that is usually 1½ to 3¼ inches wide. *Plank flooring* is also linear; however, it is wider in width. *Parquet flooring* is a series of wood flooring pieces that create a geometric design. You can get wood flooring prefinished, which is generally more durable and doesn't require the additional steps of staining and sealing.

All wood flooring is categorized in different grades. Although the grade doesn't affect the durability of the wood, it does gives it different looks. *Clear wood* is free of defects, though it may have minor imperfections, whereas a *select wood* is almost clear, but contains some natural characteristics such as knots and color variations. *Common wood* (No. 1 and No. 2) has more natural characteristics such as knots and color variations than either clear or select grades and often is chosen because of these natural features and the character they bring to a room. No. 1 Common has a variegated appearance, light and dark colors, knots, flags, and wormholes. No. 2 Common is rustic in appearance and emphasizes all wood characteristics of the species.

Grades are also characterized as *first, second*, and *third,* with price generally reduced as the grade of wood decreases.

The final consideration you have when choosing wood flooring is the cut of the board. The cut will affect the look, and the price of the boards. You can choose from the following:

- **Plainsawn.** Plainsawn is the most common cut. The board contains more variation than the other two cuts because grain patterns resulting from the growth rings are more obvious.

- **Quartersawn.** Quartersawing produces fewer board feet per log than plainsawing and is, therefore, more expensive. Quartersawn wood twists and cups less and wears more evenly.

- **Riftsawn.** Riftsawn is similar to quartersawing, but the cut is made at a slightly different angle.

Hardwood floors do need regular maintenance, including regular washing with a wood-cleaning product. You can use wood flooring in kitchens, but they may not be the best choice for bathrooms, where they will repeatedly get wet. Hardwood floors can get scratched, but they can be refinished when they start to look dull or accumulate scratches. If the refinishing is done correctly, the floor will look as pristine as new.

### Laminate
A less-expensive alternative to hardwood flooring that gives you the same look—and with none of the maintenance costs—is

laminate flooring. Laminate has increased in popularity in recent years. Crafted to look like hardwood floors, they require no maintenance and don't scratch like hardwood floors do. Some of today's brands look so much like wood that you often can't tell the difference.

Laminate flooring employs a tongue-in-groove system that is glued together but is not glued to the floor underneath, except at the joints. This is called a "floating" installation, because it "floats" above the subfloor rather than being adhered to it.

Laminate flooring is created the same way laminate counter-tops are created. A clear, waterproof, protective layer covers an image that looks like hardwood. These images come in just about every style of wood available. The core of the piece is a solid material, such as chipboard or fiberboard, which resists indentations. Most systems also have a backing layer that resists moisture penetration from the subfloor.

If your budget is tight but you like the look of hardwood floors, you may want to look at laminate.

### Ceramic or Stone Tile

Tile is a great choice for kitchens, bathrooms, foyers, and other areas that are likely to be wet or get high traffic. Ceramic tile is durable, but the tiles may crack if something heavy is dropped on them. For the most part, it's relatively easy to replace a cracked tile, so keep some extras on-hand in case of a mishap.

Tile prices can range from very inexpensive to through-the-roof! Some ceramic tiles are imported and go through extensive forming and firing processes that make them pricey. On the other hand, you can find close-outs at tile outlets at significant mark-downs, saving as much as 70 percent off retail prices.

When choosing tile, it's important to pay attention to the lots and sizes of the tiles. Be sure your tiles come from the same lot, because that will ensure the color doesn't vary. Because of the firing process involved, individual tiles can vary in size—insist on tiles of consistent size. If you're doing the tiling yourself, keep in mind that smooth tiles make grouting easier.

One other word about tile: consider the people who will be walking on it. Very smooth or slick tile can pose a slip hazard, especially for young children and older people. If you decide to use tile and have people in your household for whom slipping may be an issue, choose tiles with slightly rougher surfaces or put in safeguards, such as grab bars or area rugs with no-slip backing, to help them navigate the tile more easily.

## Carpet

Not long ago, wall-to-wall carpet was the standard in most households; in some cases, even covering beautiful hardwood floors. Its relatively low price and easy care made it an attractive option for many homeowners.

Today, with so many flooring options, carpet is still popular, with a wide range of features, quality, and pricing. The Carpet and Rug Institute (CRI) recommends looking at the style of carpet for the room you want to cover. Many innovations in carpet materials and construction have been made in recent years. The following are some of the styles and construction considerations that will affect your cost:

- **Cut pile**. In cut-pile carpeting, one of the most popular types of carpet, the loops of which the carpet is constructed are cut, leaving individual yarn tufts. The type of fiber, density of tufts, and the amount of twist in the yarn will all affect the durability of the carpet. Types of cut pile include the following:
    - **Plush/velvet.** Smooth, level surfaces, which give a formal, uniform appearance, but are easily marked by footprints.

- **Saxony.** A smooth, level finish, but pile yarns have more twist so the yarn ends are visible and create a less formal look, which minimizes footprints.

- **Friezé.** Yarns are extremely twisted, forming a "curly" textured surface. This informal look also minimizes foot prints and vacuum marks and is usually the most durable.

- **Level loop pile.** The loops in this carpet are the same height, creating an informal look. It generally lasts a long time in high-traffic areas. Many of today's popular Berber styles are level loop styles with flecks of a darker color on a lighter background.

- **Multilevel loop pile.** Usually this carpet has two to three different loop heights to create pattern effects, providing good durability and a more casual look.

- **Cut and loop pile.** This carpet features a combination of cut and looped yarns. The combination provides a variety of surface textures, including sculptured effects of squares, chevrons, swirls, etc.

According to the CRI, a carpet's fiber and construction will determine how it stands up to spills, pets, and daily traffic. Approximately 97 percent of all carpet is produced using synthetic fibers that are designed to feature style, easy maintenance, and value. There are five basic types of carpet pile fibers:

- **Nylon.** This is the most popular and represents two thirds of the pile fibers used in the United States. Wear-resistant and resilient, nylon withstands the weight and movement of furniture and provides brilliant color. It also has the ability to conceal and resist soils and stains. Nylon is generally good for all traffic areas. Solution-dyed nylon is colorfast because color is added in the fiber production.

- **Olefin (polypropylene).** Olefin is strong, resistant to wear and permanent stains, and easily cleanable. It's also notably colorfast because color is added during fiber production. Olefin resists static electricity and is often used in both indoor and outdoor installations because of its resistance to moisture and mildew. It's also used in synthetic turf for sports surfaces and in the home for patios and game rooms. Many Berbers are made of olefin.

- **Polyester.** Noted for luxurious, soft "hand" when used in thick, cut-pile textures, polyester has excellent color clarity and retention. It's also easily cleaned and resistant to water-soluble stains.

- **Acrylic.** Acrylic offers the appearance and feel of wool without the cost. It has a low static level and is moisture- and mildew-resistant. Commonly used in velvet and level-loop constructions, acrylic is also often used in bath and scatter rugs.

- **Wool.** Noted for its luxury and performance, wool is soft, has high bulk, and is available in many colors. Generally, wool is somewhat more expensive than synthetic fibers.

A wool/nylon blend combines the look and comfort of wool with the durability of nylon. Acrylic/olefin and nylon/olefin are other popular blends, offering good characteristics of each fiber.

The CRI says that when it comes to durability, there's little difference between bulked continuous filament (BCF) or staple (spun) fibers. The difference lies in the length of the fibers in the yarn, with staple having shorter lengths, which gives the yarn more bulk (sometimes described as being more like wool).

When carpet is manufactured with staple fiber, there will be initial shedding of shorter fibers. It will soon stop, depending on the amount of foot traffic and frequency of vacuuming. Wool is a naturally staple fiber; nylon and polyester can be staple or continuous filament; and olefin (polypropylene) is usually BCF.

### Saving Money on Flooring

If your budget is running tight, you can skimp on floor coverings and opt for inexpensive laminate or carpet and then upgrade to more expensive floors later on. There are many beautiful vinyl floor options on the market. Just be sure to discuss this with your plumber. The height of the flooring may affect the height at which your toilet flange is set.

The easiest way to save money on carpet is to choose the least-expensive carpet that is appropriate for the room you're covering. Wall-to-wall carpet is installed by unrolling the carpet and stapling it at the edges. Skilled installers can lay wide expanses of carpet without evidence of a seam. Carpet is laid over a pad, which adds both additional cushioning and protection from moisture from the subfloor, so you'll need to calculate that cost into your budget as well.

Carpet isn't easy to do yourself, especially when it comes to hiding seams. If you're looking for a lower-cost way to have the look of carpet without the expense, try this trick: purchase a piece of the carpet that will fit the room and comes only about 4 to 6 inches away from the walls. (Have the seller bind the edges; this usually costs about $200 or less, depending on the size of the carpet.) Lay a perimeter of laminate tile around the edge of the room, and put the carpet and pad down over the floor, overlapping the laminate. You've created the look of carpeting, with an attractive laminate finish around the edges.

When shopping for any type of flooring, look for remnants and close-outs to save additional cash.

## MONEY IN YOUR POCKET

Builders are often given discounts on myriad products. When you go shopping, wear your "work" clothes and ask for the professional rate, explaining that you're building a new home. Hank saved 50 percent off a roll of laminate flooring for a house he was remodeling because the retailer assumed he was a builder, just because of his grubby clothes and work boots. On $200 worth of flooring, that's $100! Of course, you are a builder—you're building your own home. Take advantage of the savings to which you're entitled.

**Savings for You: $100**          **Running Total: $56,130**

# Interior Doors

The doors you choose for the interior of your home will significantly impact the design of the home. You have more choices when it comes to interior doors than exterior doors, which will enable you to create a look that both suits you and makes your living space more functional.

- **Solid hinged door.** This is the traditional, solid panel. They're available in a number of types of materials, ranging from wood (most expensive) to hollow construction or composite materials or fiberglass (least expensive). Keep in mind that hollow construction doors do little to help soundproof rooms.

  When you're choosing hinged doors, you need to consider which way the door will swing. Doors that swing out mean you increase the amount of space you have in the room behind the door. However, doors that swing out are also a hazard, because the person exiting the room can't see if anyone is in the path of the door. Choose doors that swing in for most second-floor rooms and bathrooms, unless the room size or design requires otherwise.

You have a few surface choices with solid hinged doors:

- **Flat.** No surprise here—a flat door is a solid, flat surface from top to bottom. This is generally the least expensive type of door, but it also doesn't do much to add character to the home. It can be a good option for homes with a very modern design or very simple, clean lines, however.

- **Flat panel.** Flat-panel doors have recessed panels within the door. This style of door has a very old-fashioned look and can be a good choice if you're going for a vintage look.

- **Raised panel.** Raised-panel doors are more expensive than flat or flat-panel doors that have panels wider than the perimeter of the door. These panels are set in a number of different patterns, so choose the one that best complements your home.

- **French doors.** French doors are double doors that have glass inlays and that open from the middle. The doors can swing out or in, depending on the space you have. They add a great open look, even while separating rooms, but offer little in the way of privacy or soundproofing. Of course, because there are two doors and they have glass inlays, they're more expensive than traditional solid doors.

- **Bi-fold and multi-fold.** These doors are generally seen on closets, pantries, or other storage spaces. As the name implies, they open from the middle and fold—twice for bi-fold, and more for multi-fold.

- **Café doors.** Think of the swinging saloon doors in old westerns, and you've got a good idea of what café doors look like. These work well for creating a separation between two rooms that need separation but also require frequent access. These doors are frequently used in the kitchen to separate a dining room or a pantry. The back-and-forth swinging can lead to accidents, however, so don't use them in areas where there's a lot of two-way traffic.

- **Bypass doors.** Bypass, or sliding, doors are usually seen on closets. They are two doors that are designed to slide by each other on different tracks.

- **Pocket.** Pocket doors slide into a wall. Because they don't need room to swing, they're great for sealing off tight spaces. You usually need to decide if you want pocket doors in the planning phase, however, because the space needs to be framed to accommodate them.

## Saving Money on Doors

The easiest way to save money on doors is on materials. These days, composite or fiberglass doors often look like wood, so check out your options and compare pricing.

Also, to save some cash while adding a bit of flair, add molding to a flat door to give it a unique look. Be sure your door's construction can sustain the molding, and use glue—not nails—to secure molding to hollow construction doors.

Hinged doors are actually comprised of both the door and the door jamb, so you often have the choice of buying them prehung or hanging them yourself. Prehung doors are more expensive, but there's nothing more annoying or troublesome than an improperly hung door. If you're a do-it-yourselfer, be sure you have the skill set to frame the door jamb properly to ensure your door will fit securely and remain open or closed when you want it to.

## Locks and Knobs

Locks and knobs account for another door-related expense. Don't just assume that pricey knobs and locks are better.

Choosing interior door locks isn't as critical as exterior door locks, which will safeguard your home. Interior door locks are also easy to install. This is definitely an area where you can save some cash by choosing inexpensive locks and doing it yourself. No one will know the difference.

## Fireplaces

Nothing gives a space an *"Ahhh"* factor like a fireplace. The fireplace often becomes the focal point for the room, and it can also be used to cut your utility bills. You have two options when it comes to fireplaces: wood burning and gas. Wood-burning fireplaces are more expensive to install because they require a masonry box, a solid concrete or brick barrier between the studs and the fireplace that protects the wood in your home. A flue extends inside the chimney from the level of the fireplace up through the roof or wherever the fireplace is vented.

If your home uses natural gas or propane, you can save money by using a gas fireplace. Gas units typically don't require a firebox, but you do need to purchase the gas insert (ranging from $1,200 to a few thousand dollars, without covers and accessories) and, if you use propane, a propane converter (up to a few hundred bucks). Still, the net savings of a gas fireplace is significant—the venting can go directly out the wall, so you don't need to build a chimney. Some gas fireplaces don't need a vent at all, but be sure to read up on the warnings about vent-free fireplaces and air quality before you make this choice.

Another benefit of gas fireplaces is that you don't need firewood. That means no buying or cutting wood and no hauling it in the house to build a fire. Another perk: gas fireplaces turn on and off with a touch of a button.

## Look Like a Million—for Less

As you're doing the inside jobs, this far into the project, you may find that your budget is a bit stretched. If things have cost more than you expected—even with our money-saving tips—consider some of these ways to cut corners in a relatively painless way:

- **Spend visibly.** Splurge on public areas of your home. Use more expensive moldings or flooring in the foyer, living room, dining room, and kitchen, and save the less-expensive options for the bedrooms and bathrooms primarily used by family. Try similar strategies for fixtures, wallpaper, and the like.

- **Delay gratification.** Cut out nonessential things that can be done later. You don't need expensive wallpaper now—add it later. It's easy to hold off on moldings, cabinet hardware, wallpaper—even paint. The white primer that your drywall contractor applied can always be painted later.

- **Look for outlets.** Outlets, flea markets, and builder close-out sales are all great places to save money on materials.

- **Downgrade materials.** Choose a less-expensive paint or molding. Buy cheaper close-out tile and then use more expensive decorative accents to complement it, or use more expensive tile in public areas and cheaper stuff where no one will see.

## Dollar-Saving Do's and Don'ts

- Use half-walls, columns, and other partitions instead of full walls, if possible, to save on labor and materials and create a more open floor plan.

- Keep stairs simple—curves, spirals, and L shapes add cost.

- Use proper insulation to save on utility bills and add better soundproofing to your home.
- Learn about the features that add cost to interior materials and choose the least-expensive option that will do the job you require.
- Add decorative moldings to simple doors and walls to inexpensively spruce them up.
- Splurge on more expensive materials and treatments in the more highly trafficked areas of your home, and choose less-expensive options for those areas that aren't so visible.
- Delay some projects that can easily be done later to cut costs.

# 10

# KITCHENS, BATHS, AND UTILITY ROOMS

Your kitchen, bathrooms, and utility room are considered the value centers in your home. Real estate agents will often tell you that the kitchen can be a key selling point in a home, and bathrooms are another strong feature that attracts buyers—and possibly higher home prices.

That's important to remember, but these centers will also be areas where you and your family will spend a great deal of time. They are all, in one way or another, work centers, where cooking, cleaning, or personal care take place, so they need to be functional for you and your family. If you don't spend some time planning, you could end up wasting a great deal of money and feeling a mild sense of dissatisfaction every time you spend time in one of these rooms.

The good news is that you can save money, time, and effort on these rooms. We will show you how.

## Someone's in the Kitchen

How much time do you and your family spend in the kitchen? Is someone in your home a skilled cook who loves to show off his or her culinary skills? Or are you more the "if it can't be microwaved, we don't eat it" variety?

Your answers to those questions will determine whether you should spring for a gourmet kitchen or if you should outfit your kitchen space with functional but not over-the-top equipment. Take some time to think about your dream kitchen. If you don't really have a dream kitchen, perhaps you should consider planning your space with just the essentials.

## The Well-Designed Kitchen

The National Kitchen and Bath Association (NKBA) has many guidelines to help owner-builders design their kitchens for maximum productivity. Kitchen-conscious buyers will look for adherence to these guidelines when they examine a home for purchase.

### The Kitchen Triangle

Most kitchen designers will tell you that one hallmark of good kitchen design is the so-called "kitchen triangle." This simply means that your three major work centers—the refrigerator, stove, and sink—should be positioned in a triangular pattern. According to the NKBA guidelines, the basic work triangle has the primary kitchen sink opposite the refrigerator and the cooktop. In smaller kitchens that can't accommodate such a triangle, the sink should be between the refrigerator and the cooktop. Each part of the triangle should be no less than 4 feet and no more than 9 feet in length, which gives the cook room to move but doesn't let the distance between work centers inhibit the cook's productivity.

### Doorways

The clear opening of a doorway should be at least 32 inches
wide. This would require a minimum 2-foot, 10-inch door. No
entry door should interfere with the safe operation of appliances,
nor should appliance doors interfere with one another.

### Work Stations

When the kitchen plan includes more than three primary
appliance/work centers (the "kitchen triangle"), each additional
travel distance to another appliance/work center should measure
no less than 4 feet and no more than 9 feet from the center
front of the appliance or area to the center front of the next
nearest appliance or area. These additional work centers may
include a food preparation area, which is usually a countertop;
a food storage area, often cabinets or a pantry; a dishwasher; or
a microwave oven.

If you have an island in your kitchen, the NKBA recommends
that no work triangle leg intersect an island/peninsula or other
obstacle by more than a foot and that no full-height, full-depth,
tall obstacles, such as pantry cabinets, refrigerators, or oven cabinets, block the flow of traffic between two primary work centers.

### Walkways

It's recommended that you allow at least 42-inch aisles for
kitchens with one cook and at least 48 inches if you'll have two
or more cooks working together regularly. In any case, the width
of the walkways should be no less than 3 feet wide.

### Seating Clearance

Any kitchen traffic should pass at least 32 to 44 inches from the
nearest dining seat. You should also pay attention to kitchen

seating areas and their clearances, such as those for your kitchen table or for seating at an island or breakfast bar:

- **For 30-inch-high tables and counters.** Allow a 24-inch-wide by 18-inch-deep counter/table space for each seated diner.

- **For 36-inch-high counters.** Allow a 24-inch-wide by 15-inch-deep counter space for each seated diner and at least 15 inches of clear knee space.

- **For 42-inch-high counters.** Allow a 24-inch-wide by 12-inch-deep counter space for each seated diner and 12 inches of clear knee space.

### Food-Prep Areas

Be sure you have plenty of space for food preparation. The NKBA calls the prime prep counter space—that stretch of counter adjacent to a sink or appliance—the *landing area*. Your landing area should be at least 2 feet wide on one side of the sink, 18 inches on the other, and at least 16 inches deep. The best-case scenario is a 3-foot-wide and 2-foot-deep space next to the sink.

### Cooktops and Ovens

Plan for a minimum of 12 inches of landing area on one side of a cooking surface and 15 inches on the other side. If the cooking surface is at a different countertop height than the rest of the kitchen, the 12- and 15-inch landing areas must be at the same height as the cooking surface. Your municipality may have building code restrictions about cooktop placement, so be sure to follow those guidelines, as well as those of the appliance manufacturer.

Never place a cooktop under an operating window, where window treatments can cause a fire hazard. Reconsider putting your cooktop within an island, where little hands can reach it from the other side. If you must locate your cooktop there, and if the counter height is the same as the surface-cooking appliance, extend the countertop a minimum of 9 inches behind the cooking surface.

Your cooktop will need to be properly ventilated, as well. Be sure to follow the manufacturer's recommendations, as well as your local building codes.

If your cooktop is separate from your oven, you'll likely also have a wall oven. These can be single or double and are housed in a cabinet designed for ovens. Follow the manufacturer's recommendations for oven placement, and be sure no doors intersect the arc of the oven doors when they open. The NKBA also recommends a 15-inch landing area adjacent to the oven.

## Refrigerators
Your refrigerator should be placed flush against the wall, and it's recommended that you include at least one of the following:

- 15 inches of landing area on the handle side of the refrigerator
- 15 inches of landing area on either side of a side-by-side refrigerator
- 15 inches of landing area that is no more than 48 inches across from the front of the refrigerator
- 15 inches of landing area above or adjacent to any under-the-counter-style refrigeration appliance

## Dishwashers
Minimize your time handling dirty dishes by keeping your dishwasher within 3 feet of the nearest sink. You should have at least 21 inches of space between the edge of the dishwasher and any counters or appliances that are placed at a right angle to the dishwasher.

## Microwaves

Microwave ovens can sit on a kitchen countertop, or you can get a model that has brackets that mount the microwave beneath a cabinet. Some cabinets come with a built-in shelf especially for microwaves. The type and placement of your microwave depends completely on your preferences. Something to keep in mind: although counter-fastening models and cabinets with special shelving free up counter space, they also cost more.

The NKBA recommends that the bottom of the microwave be 3 inches below the principle user's shoulder but no more than 54 inches above the floor. In any case, it should be no less than 15 inches off the finished floor with at least 15 inches of landing area above, below, or adjacent to the microwave's handle side.

## Counter Space

With all these landing areas, it may seem as if you're going to need a really big kitchen. Not really—if two landing areas are adjacent to one another, determine a new minimum for the two adjoining spaces by taking the longer of the two landing area requirements and adding a foot.

Overall, the NKBA recommends a total of 158 inches of countertop frontage, 2 feet deep, with at least 15 inches of clearance above. Countertops may have clipped or round edges, which are preferable to sharp edges.

## GO FIGURE

The NKBA formula for determining shelf and drawer frontage is determined by multiplying the cabinet size by the number and depth of the shelves or drawers in the cabinet, using the following formula:

| Cabinet width in inches | × | Number of shelves/ drawers | × | Cabinet depth in feet (or fraction thereof) | = | Shelf/ drawer frontage |

### Lighting

Your municipality will have codes about lighting, but you should think about the light you'll need for your workspaces as well. The NKBA recommends that window and skylight area equal at least 8 percent of the total square footage of the kitchen, or a total living space, which includes the kitchen.

### Garbage Containers

Of course, you're going to need a waste bin or two for the kitchen. The NKBA recommends at least two: one at each end of the cleanup/prep sink and a second for recycling. The latter can be located either in the kitchen or nearby.

## Your Dream Kitchen

When you have the specs for your dream kitchen in place, now comes the fun part—picking out the elements that will make your kitchen the heart of your home. We discussed flooring options already in Chapter 9, so you need to make decisions about spending or splurging on four primary areas: appliances, countertops, cabinets, and sinks and fixtures.

### Appliances

In general, you're going to need at least two appliances in your kitchen—and probably more like five or six. At the bare minimum, you need a refrigerator and a cooktop/oven combination.

In addition, many new kitchens have a dishwasher, microwave, and garbage disposal. And for the really high end, you may want to add double-wall ovens, an additional freezer, a small under-cabinet refrigerator, and a wine cooler.

### Saving Money on Appliances

The key to saving money on appliances is to shop around. Between sales and pricing differentials with various retailers, you can save hundreds or thousands of dollars just by taking advantage of seasonal opportunities. Shopping online can also be a good way to compare prices. However, be sure to check out shipping charges, which can wipe out any savings you might enjoy through an online retailer.

Before making any major appliance purchase, it's a good idea to check *Consumer Reports,* which can be found at www.consumerreports.com or at your local bookstore. This independent, not-for-profit organization road-tests appliances, as well as hundreds of other big- and small-ticket purchases, and rates them for quality, value, durability, and other factors. There is a nominal fee for one-time access to the service, or you can purchase a yearly subscription for about $26.

At independent appliance retailers, prices are often flexible, so don't be afraid to use your negotiating skills. Ask the retailer for a discount if you order multiple appliances at the same time. Such a large order has more profit for the retailer, so he or she may be willing to knock 10 percent or so off the purchase price.

## MONEY IN YOUR POCKET

For savings on appliances, look for discontinued models. Retailers often mark down older models, many of which are still brand new, to make way for newer versions from the manufacturer.

You can also try scratch-and-dent retailers, who specialize in merchandise that's been slightly damaged. Often, if the scratch or dent is on the side or back of the unit, it will be concealed by the surrounding cabinets.

We've seen mark-downs as much as 40 percent in these cases. On a $1,000 appliance, that's $400.

**Savings for You: $400**          **Running Total: $56,530**

Also don't forget the energy-efficiency factor to save money. The American Council for an Energy Efficient Economy estimates that appliances can be as much as 10 percent of a home's energy bill. EnergySTAR estimates that a washer with the EnergySTAR label can save you up to $120 per year on your utility bills because they typically use 50 percent less energy and almost half the water as other machines.

**DON'T TRIP**
on your **SHOESTRINGS**

Some high-end appliances, such as certain brands of ovens or refrigerators with built-in water and ice dispensers, may need special gas or water hookups. If you plan on having these high-end appliances in your kitchen, make that decision before the plumbing and HVAC are done so the appropriate professional can make accommodations for them. Your contractor will also need to know your cooktop layout to install a vent for it. If you don't account for these needs up front, you may find yourself racking up additional fees to have the plumber or technician add the connections later.

## Cabinets

The cabinets will likely be the largest part of your kitchen budget. When you're choosing your cabinets, you have three choices:

- **Stock cabinets.** Choosing stock cabinets can be a far less expensive option than semi-custom or custom cabinets. These manufactured cabinets come in standard sizes with a limited variety of colors and options. Sizes usually increase in 3-inch increments, and fillers are used to cover gaps. Stock cabinets may be available at some retailers to take home the same day. In the worst case, they can be ordered and delivered in a few weeks.

- **Semi-custom cabinets.** Semi-custom cabinets are built at the manufacturer, based on your kitchen's measurements. They are usually about twice the price of stock cabinets, but they usually have more features and styles available. Sizes are usually based on the same parameters as stock cabinets, however, so they're not as individually tailored as custom cabinets. They require a longer lead time to manufacture and install than stock cabinets, so it's a good idea to begin working with the provider at least 4 months ahead of time or as soon as your sheet rock is installed.

- **Custom cabinets.** Custom cabinets are the cream of the crop. Built specifically to your home's measurements, custom cabinets have no gaps to fill or spaces to cover. They are usually built and finished on-site by a carpenter. They are the most expensive cabinet options, but they offer you better use of space. They also require the longest lead time to install, so it's a good idea to work with the contractor at least 6 months in advance of your desired installation date.

## MONEY IN YOUR POCKET

If you're flexible about the cabinetry you choose and have a bit of time, check various kitchen and bath showrooms, including supercenters, which change their displays regularly. We know a family who got $15,000 worth of cabinetry and countertops for $2,000. If you need to supplement the cabinets, you can order the units you need.

**Savings for You: $13,000**          **Running Total: $69,530**

Beyond the general look and size of your cabinets, you also need to consider several other features, which will add to your cost.

**Construction** The least-expensive option in stock and semi-custom cabinets is a particle board construction, which is essentially sawdust and glue covered by a laminate. Although that may sound like it won't hold up well, these basic cabinets can be quite sturdy. The concern with these cabinets is usually the hinge attachment, which can fail when the flake board eventually breaks down.

To solve this issue, some manufacturers have constructed the back and sides of the cabinet from particle board and the door and cabinet front from hardwood. In these combination cabinets, the hinge attaches both to the hardwood door and the hardwood front, making the hinges much more durable.

Overall, full wood construction is considered the most durable and the most desirable, but that can add between 5 and 20 percent to your cost, depending on the cabinets you order.

**Overlay** Overlay refers to where and how much the cabinet door sits on the cabinet front. Full overlay means you see mostly cabinet door and little cabinet front. Partial overlay, which uses less wood, means the cabinet door covers little more than the cabinet opening, leaving more of the cabinet front exposed between the doors. Partial overlay is generally less expensive.

**Interior Fittings** Cabinets have come a long way in recent years, and many are available with a wide range of interior options, such as built-in utensil holders, lazy Susans, pull-out drawers, and the like. In addition, many have soft-shut functions that prevent the doors from slamming shut.

Some of these options are simply convenience; others, such as pull-out drawers, which help save time when searching for something that's in the back of the cabinet, are terrific time-savers. Most of these interior fittings, however, will add to your cost. Custom and semi-custom cabinet retailers will likely have a price list of the various options.

**Hardware** Most cabinets don't come with handles on them. You must choose from a variety of hardware options. These can become pricey, so shop carefully for them. You don't need to buy your hardware from the same place you bought your cabinets, so look around for the best price.

## MONEY IN YOUR POCKET

If you have a bit of carpentry ability, you can save quite a bit of money by installing cabinets yourself. A recent *Consumer Reports* project reported cabinetry installation for a 10×10 kitchen at approximately $2,500.

**Savings for You: $2,500**      **Running Total: $72,030**

### Countertops

Your countertops will be your kitchen workspace and should complement your cabinet choices. When it comes to choosing countertops, you have a variety of options, including the following:

- **Laminate.** These countertops are the most affordable and have a plastic barrier on the top that generally makes them easy to clean. Because these countertops are made of layers of material different from the surface, the edge of the countertop may be a different color than the surface and needs to be covered or fashioned to hide that difference. Some people refer to laminate countertops as "Formica," after an early brand of these plastic laminates.
- **Solid composites.** These solid-surface materials (a commonly known brand is Corian) are generally more durable than plastic laminate countertops, but they're also more

expensive and can cost as much as stone. Because they're a solid surface, however, scratches, cuts, and burns can usually be buffed out of the material. Again, because it's solid, you have different edging options than traditional plastic laminates.

- **Wood.** Hardwood countertops are usually constructed of pieces of wood glued together. Often called "butcher block" counters, this is a moderately priced option (more expensive than laminate, less expensive than stone or solid composites) and can give a kitchen a traditional look.

- **Tile.** Tile countertops can be an inexpensive option that lets you add color and uniqueness to your kitchen. Depending on the tile you choose, you can install it inexpensively, and it's generally not affected by heat. However, dropping something heavy on the tile may cause it to break, in which case, you'll need to replace the cracked tile. When you buy your tile, keep some extra on hand just in case.

- **Stone.** Hard stones, such as granite, make beautiful countertops that are heat-resistant, cut-resistant, and will even stand up to a dropped pot or two. More brittle stones, such as marble, make less-durable countertops, but they are equally as beautiful. Stone is the most expensive option—granite can run $50 or more per linear foot—but with proper care, is likely to last as long as you own your home.

## Sinks and Fixtures
You may think of your kitchen sink and its fixtures as secondary, but your sink will be one of your primary workplaces. You have several sink options: single or double basin, a flush-with-the-counter sink, or one with a rim that sits on top.

You might have a bit of sticker shock when you see the price of some fixtures, which can run as high as $1,000 or more for a faucet! Think about whether you want hot and cold knobs or whether you want a single lever that manipulates water flow and temperature. Some faucets also integrate a spray mechanism, whereas others have a separate spray mechanism.

Sinks and fixtures come in a variety of materials and styles. It's possible to find sinks and fixtures as low as $100 to $200. Be especially careful of manufacturers' sink measurements: they're usually the inside of the bowl, whereas you need to measure the outside to be sure the countertop opening is cut properly. Otherwise, you or your contractor could end up ruining your countertop.

### Saving Money on Your Kitchen

When you purchase countertops, especially laminates and composites, be sure to have your exact measurements. Ordering countertops that are too short will likely mean they can't be used, so you'll need to reorder—at more than double the expense.

Some countertops lend themselves to do-it-yourself installation—tile and some plastic laminates, for example—but others, such as stone and wood, are more difficult to install should be handled by a professional.

Kitchen designers we've talked to claim that the two biggest wastes of time and money in the kitchen are related to planning and measurements. When people don't take time up front to plan their kitchens, such as appliance and countertop placement, they realize after the installation process begins that appliances and work areas are inconveniently placed and need to be reworked, adding cost. As one designer told us, you should take some time to "cook a meal in your mind." Walk through the process of food prep in your layout, and try to think through the best design for your style. For instance, if you're left-handed, you may want different placements than those you might see for right-handed cooks. This generally won't affect resale value, but it's a good idea to still stick to basic principles, such as the triangle layout, to make your kitchen most functional.

The other big waste of time and money is not having exact measurements. It's common for owner-builders to place cabinet or countertop orders based on blueprint measurements. However, the actual house measurements may vary by as much as a few inches. When the materials arrive, they're the wrong size and either need to be reordered or refitted at more expense to the homeowner. Wait until the room has sheetrock before you order your cabinets and counters; then take careful measurements or have the retailer or contractor come out and take measurements to save you time and money.

## MONEY IN YOUR POCKET

Why pay for kitchen design when you can get it for free? Some cabinet retailers and kitchen designers may charge you $500 or more for a layout, but many home centers will do it for free to get your business. Usually you need to make an appointment and have your measurements in hand. You also need to pick out a brand of cabinet from their offerings, because their design programs are brand-specific. However, this will give you a good idea of the cost of your cabinets and the overall layout of your kitchen, even if you choose a different provider. Many of these programs also point out problematic placements, such as when two doors bump into each other.

**Savings for You: $500**          **Running Total: $72,530**

# Take a Bath

Bathrooms are also important sales centers for your home. Plus, the right bathroom can add enjoyment to your life. Think about relaxing in a steam shower or a whirlpool tub—or simply having an extra bathroom to eliminate family arguments over morning mirror time.

You want to plan your bathroom design based on the room in your house. For instance, your master bathroom will likely be the most elaborate, whereas near your family room, you may just need a half-bath with a toilet and sink. You can design a home so every bedroom has a bathroom or so family members share a common bath. Some home designs have a bathroom that connects between two rooms, often called a Jack and Jill bathroom. The number and design of your bathrooms is up to you.

# The Well-Designed Bathroom

The NKBA has terrific guidelines to help you create bathrooms that are functional as well as stylish.

### General Structure

The doorway to any bathroom should be at least 2 feet wide with no other door or fixture intersecting its path. Plan a clear floor space of at least 30 inches from the front edge of all fixtures (lavatory, toilet, bidet, tub, and shower) to any opposite bath fixture, wall, or obstacle.

Ceilings should have a minimum floor-to-ceiling height of 80 inches over the fixture and at the front clearance area for fixtures. A shower or tub equipped with a shower head should have a minimum floor-to-ceiling height of 80 inches above a minimum area and 30×30 inches at the shower head.

## Toilet

The center of the toilet should be at least 20 inches from any wall or cabinet side, with a minimum of 15 inches, and with a height of between 32 and 43 inches to fit the user. In larger bathrooms, consider putting the toilet in a separate smaller room.

## Shower and Tub

The NKBA recommends that the interior for a standalone shower be at least 3 square feet, or a minimum of 30 inches by 30 inches. A seat within the shower is also recommended and should measure 17 to 19 inches above the shower floor and 15 inches deep.

Users should be able to access shower controls from both inside and outside the shower spray or tub and be between 38 and 48 inches above the floor, according to the user's height. They should be either pressure-balanced or thermostatic mixing. Both features regulate water temperature so, for instance, if someone flushes the toilet, it doesn't change the water temperature in the shower.

Of course, the area around the shower or tub should be waterproofed with either tile or a tub surround that's at least 72 inches above the floor, and grab bars are a good idea to help users navigate slippery floor surfaces when getting in and out of the shower or tub.

Any glass used to enclose the tub or shower or installed on a hinged shower door must be tempered or an approved equal, and it's not recommended that any steps be placed outside the tub or shower, because they get slippery and could cause a fall.

## Accessories

Mirrors should be above or near the lavatory at a height appropriate for the user. Mirrors come in many sizes and shapes.

A current trend installs mirrors flush against the wall rather than covering a cabinet. We recommend that you examine your storage space and determine whether you can afford to sacrifice that storage. Such "medicine cabinets" are great places to store toiletries and hygiene products, so think carefully before choosing a flush-to-the-wall mirror.

The NKBA recommends that the toilet paper holder be located 8 to 12 inches in front of the edge of the toilet bowl, centered at 26 inches above the floor. Additional accessories, such as towel holders, soap dishes, etc., should be conveniently located near all bath fixtures.

Lights should be provided for each functional area of the bathroom, with at least one wall switch. Don't locate hanging fixtures within 3 feet of the tub or shower perimeter.

### Heat

Because bathrooms can get cold, consider a supplemental heating source, such as a heat lamp, toe-kick heater, or even radiant floor heat. The NKBA recommends a minimum room temperature of 68°F.

## Bathroom Tactics

Even if you have several bathrooms in your house, you won't want to treat them all the same. You need to examine how many people will be using them, as well as what those individual users' needs are. The following will be your biggest bathroom expenses:

- Bathtub/shower combination or separate bath and shower
- Sink
- Cabinet
- Countertop
- Tub surround/tile and enclosure
- Fixtures

You could also include a bidet, a steam shower, or additional cabinets/shelving.

### Get Your Mind *in* the Toilet

You might have never given much thought to toilets, but you will when you're planning your bathroom. They come in a range of styles, colors, and shapes, with standard bowls being about 14 inches. For larger users, larger models are available. Toilets are usually made of porcelain, and the color you choose is up to you. White units are typically less expensive and are the most versatile for future bathroom redesigns. We don't recommend using trendy colors, because they'll likely look dated in a few years. Incorporate trends with paint, wallpaper, or accessories that you can easily change.

Toilets are constructed as one-piece units (which are more expensive) or two-piece units that have a base and a tank. Old-fashioned wall-mounted toilets have a tank attached to the wall above the user's head-level when seated.

Some homeowners opt for a bidet as well. This should be matched to the toilet design and positioned next to the toilet, with the center of the bidet a minimum of 16 inches from the center of the toilet. Either way, choose water-conserving varieties to save money on utility bills.

### Tubs and Showers

You're going to wash the cares of the day away, or start your day fresh, in your tub and shower. In addition to being an essential hygiene facility, your tub or shower can greatly add to your quality of life.

You have two options when you choose your bath and shower setup: you can choose to have a bathtub with a shower over it, or you can choose to have separate bath and shower units.

Before you decide to have separate tub/ shower units, however, consult your local tax department. Your taxes may be affected by the number of fixtures you have in your bathroom.

Bathtubs come in a variety of styles. Standard recessed tubs are what you most often see in bathrooms today. Finished on three sides, this style of tub fits flush against the wall and is generally the least expensive option. Free-standing tubs come in footed, often called claw-foot, or pedestal styles. Drop-in tubs are unfinished around the sides and fit into a platform, much like a large sink. For extra luxury, you can choose a whirlpool tub that circulates water through jets embedded in the tub's sides.

## Sinks

You have more decisions to make about bathroom sinks than kitchen sinks. For instance, for small spaces, you can opt for a pedestal sink—one that sits on a narrow base. A more traditional option is a sink mounted on top of a cabinet, which provides greater storage space. Some sinks are mounted directly to the wall and may be skirted, affording some limited storage space underneath. Some bathrooms have two sinks, especially those in master bathrooms or bathrooms that connect two rooms. What you choose will be based on the configuration of your bathroom and the space you have available.

## Fixtures

You have more fixture considerations in bathrooms, too. You have to decide on shower and sink fixtures and sometimes toilet fixtures. Again, most shower fixtures need some type of temperature-regulation system as a safety feature.

How do you get great deals on fixtures? Here are a few tips:

- **Look for discontinued models.** Like most other things, fixture styles are updated periodically. Discontinued models may be discounted, so check with plumbing supply houses or supply retailers.
- **Check online.** You can often enjoy savings from online retailers. Shop around on the Internet—including eBay,

where you can find new products as well as used items. You may be surprised at the savings you'll find.

- **Buy floor models.** The bathroom floor displays at retailers and plumbing supply stores are fitted with fixtures. Check and see when the designs will be changed, and you could save more than half of the cost by buying the floor model fixtures.

### Saving Money on Your Bathroom

There are a number of other ways to save money when designing your bathrooms:

- **Do it yourself.** Installing simple sink fixtures or vanity cabinets is an easy way to save installation charges.
- **Combine fixtures.** Bathtub/shower combinations are less expensive than separate units and may save you money on your taxes.
- **Heat it up.** Instead of installing extra heating units, if needed use a small space heater to take seasonal chills out of the bathroom or replace the bulbs in some of your fixtures with heat-emitting bulbs.

## It's All About Utility

Utility rooms are another work station in your home. This is where the washer and dryer are usually located, as well as home cleaning items. Depending on the layout of your house, this may also be where heating units or fuse boxes are located.

When it comes to utility rooms, your first consideration should be placement. Some homes have utility rooms located in the basement, but it's critical that you examine how your utility room placement will affect your work. Do you really want to haul laundry up and down stairs? Does it make sense to have a chute from an upper floor so you can send laundry directly to the utility room? Think about how you can use your utility room to simplify work.

With the range of cabinets and appliances available today, utility rooms don't have to be ugly. Use the same design philosophies you have for your kitchen and bathroom to make them most functional:

- **Appliance placement.** You can place the washer and dryer side by side, or you can find stacked models with the dryer on top of the washer. These appliances come with a variety of features and at various costs. The more bells and whistles you choose, the higher the cost.

- **Storage.** The utility room is one place where storage is key. Think about where you will store detergent, mops, brooms, the ironing board, and other items that will be housed in this room, and choose cabinets or storage units that will accommodate them.

- **Sink.** If you have room, it's a good idea to add an extra sink in the utility room for messy jobs, such as soaking stained clothing or rinsing mops. The sink can be an inexpensive stainless-steel model, which will be durable.

Again, planning is essential to creating a utility room that will make doing housework more convenient. Who knows? With the right fixtures, cabinets, and design, you may even enjoy housework!

## Dollar-Saving Do's and Don'ts

- Enhance resale value by using accepted standards for kitchen and bathroom design.
- Take advantage of scratch-and-dent outlets and sales for appliance discounts.

- Check out opportunities to purchase floor models.
- Save money by taking on some projects yourself.
- Combining fixtures, such as shower/bath combos, saves money on fixtures and taxes.

# 11

# LANDSCAPING, DRIVEWAYS, AND EXTERIOR LIVING SPACES

After your beautiful new home is complete, you still have a few more details to finish. Look out the window. Shouldn't the exterior of your home provide a beautiful view? What's the fun of looking through those crystal clear windows if all you see is mud, sand, and a few leftover scraps of plywood?

Paying attention to your exterior living space not only enhances your interior by providing beautiful views, but also adds to your amount of living space and makes your property more functional for your family's needs. Proper landscaping can make your home more energy efficient (really!). Decks, porches, patios, and the like add to your living space by giving you another area to entertain, dine, and relax when the weather is appropriate. Not to mention that nice landscaping—and the "curb appeal" (the level of appeal your home has to passers-by) it delivers—can add to the value and resale potential of your home.

In addition to landscaping and outdoor living spaces, you first need to determine proper access to your home through a driveway and access paths.

## Driving Home

Don't discount your driveway when it comes to planning your home. The length of your driveway will affect your overall cost—the farther away your house is from the road, the more it will cost to put in a driveway and run utilities to your home.

Some people look at a driveway as more than just an entry to the house. A paved or concrete area adjacent to the house can be a recreation area, great for playing basketball, as well as an area for children to use bicycles, skateboards, and roller skates. These usage considerations will affect which of the several driveway materials you might want to use.

### Asphalt

Asphalt, a hardened surface comprised of tar and stone, is one of the most popular driveway choices. It's durable and smooth, yet less expensive than concrete. Asphalt does need maintenance, however, and requires regular sealing. Sometimes cracks need to be repaired, and it may need to be resurfaced every 10 years or so. It can get soft in very hot temperatures, too.

### Gravel

Gravel is about the least-expensive option you can choose for driveway surfacing. These small stones are poured onto the driveway's surface, sometimes over a fabric barrier to prevent them from sinking into the soil below. Stones can be round or more square or triangular, but shouldn't have very sharp edges, for obvious reasons. You can usually install a gravel driveway yourself, saving up to a few hundred dollars in installation charges.

The downside of having a gravel drive is that the stones will likely drift out of your driveway—they'll be carried away in vehicle tires, they'll spill out into the street, etc. And if you've got very heavy vehicles sitting on them or high traffic in your driveway, they're likely to sink into ruts. The gravel will need to be replenished regularly. Putting small barriers (logs, railroad ties, decorative retaining wall, etc.) on the sides of the driveway helps contain the gravel.

## Concrete

Concrete is a more expensive option than asphalt, but it doesn't require the long-term maintenance. Concrete is a sturdy, durable surface that can pretty much be poured and forgotten. If a crack develops, it can be patched easily.

Today's concrete comes in a great variety of colors and can also be stamped to look like paving stones or other designs at less cost than using actual paving stones.

## Paving Stones

Paving stones, also called pavers, are interlocking pieces of brick or concrete that make great-looking driveways, but they're just about the priciest option available, with stones running $5 or more each. Multiply that by the 2,000 or so stones you may need to pave a driveway, and you're looking at an expensive option—but one that has dramatic curb appeal.

## Planning Your Driveway

Depending on where your track pad was placed, you'll likely have an idea of where your driveway will be placed. Before you finalize it, however, you should consider a few factors in your plan:

- **Width.** If your driveway extends from the street to your garage, it needs to be at least as wide as the number of bays your garage has. So if you have a two-car garage, your driveway will be wide enough to meet both bays, with a foot or more of additional width on either side. (Check your local building codes to see if there are specific requirements.) However, you may want to narrow the width as you get toward the street, depending on the number of cars that will regularly be using the driveway, or the other vehicles, such as boats, trailers, recreational vehicles, etc., that will be parked there.

- **Shape.** The least-expensive type of driveway is a straight run from the street directly to your garage. However, you may want to add a small lot to the shape, to allow cars to back up and turn around or to allow a space for recreation. Another convenient feature is a horseshoe or semicircle driveway. Both of these features allow cars to enter the street without backing up the entire length of the drive, which is somewhat safer, especially if children are around.

- **Exit point.** Take a look at the exit-point options. This is the point where your driveway meets the road. If your lot is situated on a curve, you may need to move the exit point to the safest location available to you, or at the very least install mirrors across the street so you can see oncoming vehicles. Work with your municipality to determine how to make a so-called "blind driveway" safer.

- **Slope.** The slope of your driveway is also a consideration. A steep driveway can be dangerous, especially in snow and ice, and, if sloped toward the house, can act as a literal waterfall. If you find that your driveway is too sloped, work with your excavator to make it more level.

If you live in an area that gets snow, remember that your driveway will need to be shoveled, plowed, or otherwise cleared. This can end up being a big expense and very labor-intensive.

In most installations, the driveway trench is dug out about 6 to 8 inches and a layer of gravel is poured. Then the surface material is applied, or in the case of a gravel drive, more gravel is laid.

Again, the key areas to save money on your driveway are the length and the material. Side-entrance garages require a bigger driveway expense because the driveway needs to extend along the side of your home and must have enough width to allow cars exiting the garage to back up properly. As discussed in Chapter 5, that expense can be eliminated with a front-entrance garage.

If your driveway is made of a solid material, you'll also want to be sure it's slightly sloped on either side to provide for water runoff. Driveways should sit slightly above the surrounding ground so water does not pool in them, which can cause additional wear and tear.

## MONEY IN YOUR POCKET

As you research the services you need to purchase to complete your home (roofing, landscaping, paving, etc.), ask your new neighbors if any of them need the same services. By lining up two or three neighbors who need their roofs repaired, driveways paved, or landscaping done, you could negotiate with the contractor for as much as 20 percent off a job. Because the contractor can do the jobs in quick succession, possibly using the same resources he uses for your project, he gains more profit, and he might be more willing to negotiate the price. If you are able to line up three landscaping jobs and knock 20 percent off your $2,000 bill, that's $400 savings for you.

**Savings for You: $400**          **Running Total: $72,930**

# Walkways and Paths

In addition to your driveway, you'll need walkways and paths to access various areas of your home, such as the route from your driveway to the front door and perhaps along the side of your home. This will keep visitors from walking on your lawn or landscaping and also give them a more comfortable and clear direction from which to enter your home. It will keep your home cleaner, too, by helping people avoid mud, sand, soil, and moisture that can be carried in on their shoes.

These walkways can be created out of paving stones, concrete, or gravel (just be sure it's easy to walk on). You can also get creative and use large flat stones, brick, wood, laminate decking products, or other material to define where you want people to walk.

Before you decide on a material, consider the people who will be using these pathways. If you have young children, elderly people, or individuals with disabilities using the walkways on a regular basis, it's probably best to keep them smooth and simple, using concrete or pavers. Otherwise, create your paths using your imagination as your guide!

# Steps to Inside

Because the entrance to most homes sits well above ground level, you'll likely need steps to enter at least one area of your home. Exterior steps are constructed similarly to interior steps— uniform treads and risers being essential for safety.

Exterior stairs need to be crafted of materials that can weather the elements—and that don't get very slippery. These generally include concrete, brick, pressure-treated wood, and laminate decking material. Some companies even make prefabricated stairs of pressure-treated wood covered by a slip-resistant, fiberglass coating.

Remember the principles to determine the slope of your stairs: the ratio of the total rise to the total run gives you the slope of the stairway, with a goal of about 30 to 35 degrees in slope. Again, check your local building code requirements.

The width of your exterior stairs is determined by the width of the entrance to the home, as well as local building codes. Also, some local building codes require that front stairs or entryways have a pediment or awning above them. If this is required and will need to be supported by the stairs, you have to add width to accommodate them.

The best way to save money on exterior stairs is to keep them simple—straight-run stairs are less expensive than curved—and to keep balusters and railings straightforward. Compare the cost of materials as well—pressure-treated wood stairs cost less than concrete stairs, but they may also need to be replaced in several years and may not add the curb appeal that you want. If you want to cut costs, put the best stairs you can in the front of your house and use less-expensive materials in the back, where they aren't as visible.

## Ramping Up

If someone living in your home uses a walker or wheelchair or has difficulty navigating stairs, you'll need an access ramp. According to the Americans with Disabilities Act, passed in 1990, the slope of the ramp should be no more than 12 inches of ramp for every 1 inch change in grade. So if the slope to your front door is 21 inches, that means you should have 21 feet of ramp to make it easy to access. The ramp can be a straight run (least expensive) or can be constructed in a zigzag manner that takes up less space. Keep in mind, however, that the zigzag type is more difficult to navigate.

Most ramps are concrete, although some are made of pressure-treated wood or laminate decking material. The ramp should be sufficiently wide enough to navigate—a minimum of 48 inches is recommended. Also any ramp that is more than 6 inches in height requires handrails.

Combine the ramp installation with that of your foundation, exterior stairs, or concrete walkway. By booking multiple jobs with the same contractor, you'll likely be able to negotiate a discount of at least 10 percent.

## Exterior Living Spaces

Who hasn't dreamed of sipping lemonade while sitting on a beautiful backyard patio? Or having a big family barbecue on a brand-new deck? These house features do more than just add appeal; they expand your home's amount of living space on a seasonal basis.

### Decks

Decks are usually very popular extensions of the home, offering the same additional dining, entertaining, and general-use space. Because they are a more complex construction project, however, they're usually more expensive to build.

When you start to plan your deck, again think about how it will be used. Will you have a grill, dining chairs, or chairs for sunbathing? Will children be playing on the deck? What may look like a broad expanse of deck soon becomes a much smaller space when you add furniture and other lifestyle items. Consider some of these items as well:

- **Location.** Most decks are built to adjoin the home, with a door from the back of the home that leads directly onto the deck. That style has its benefits, but if your home doesn't lend itself to that or if you have a better deck location somewhere else in your yard, don't limit yourself. A free-standing deck in the middle of a backyard garden or next to a swimming pool can be a refreshing change of pace.

- **Slope.** As when building your home, you should consider the slope of your backyard. If you're dealing with a significant slope, you may be able to use a stepped deck—a deck that has different levels—to make the space more functional.

- **Height.** The height of your deck will impact your cost— the higher the deck, generally, the higher the cost because of the additional decking material and labor involved.

- **Size and shape.** As with any building project, the more rectangular your deck is, the less expensive it will be to build. Octagons, circles, and other interesting angles look great, but they require more labor and will generate more waste of materials.

- **Stairs and railings.** Whether your deck adjoins to the back of your home or is free-standing, you'll need to have stairs and railings that are safe and meet the code requirements of your town. Be sure stairs are uniform and railings are well-secured and will not buckle or wobble under the weight of someone using them for support.

- **Built-in features.** If you're really ready to go all-out on your deck, the sky's the limit. Some people build in spas or hot tubs, grills, or other elaborate features in their deck. You can build in benches or storage bins, as well. Your only limitation is your budget.

After you've designed your deck, you need to choose the materials with which it will be built. Obviously, the deck is an outdoor structure, so you need to use materials that will not easily decay when exposed to the elements, insects, and decay.

Even though CCA pressure-treated wood was taken off the market, retailers are still allowed to sell through their stock, so there may still be some out there. Be sure to ask what the wood is treated with; if it's CCA, make another selection.

**Pressure-Treated Lumber**   Some people use traditional lumber for decking, but it really doesn't hold up that well when exposed to extreme temperatures, even when it's been treated and sealed. Pressure-treated lumber is a better option.

Pressure-treated lumber got a lot of media attention in January 2004, when the Environmental Protection Agency (EPA) banned the use of chromated copper arsenate (CCA) as a preservative in wood that was slated for residential use, except in permanent wood foundations. The concern stemmed from the fact that CCA is a carcinogen. However, a new breed of pressure-treated wood cropped up, which uses more copper, thus making it safer to use but at the same time boosting the price of wood as much as 35 percent. That still doesn't make it as expensive as other materials, though.

Sealing the wood every few years will preserve it longer, and pressure-treated wood holds up well against insects and weather while resisting mildew. The wood can also be stained a variety of colors.

**Composite Material**   Many people have dubbed composite decking material "Trex" after one of the most popular brands, but there are actually several on the market. Composite material will appeal to your inner environmentalist because, depending on the brand, it uses a combination of sawdust and everything from recycled milk jugs to plastic grocery bags to scrap vinyl products. It's sturdy, maintenance-free, and comes in a variety of colors. Some critics of composite material claim it mildews. That can be solved by washing the deck regularly or spraying the mildewed areas with a bleach compound.

**Vinyl Decking**   Even though this decking product is more expensive than other options by about 20 to 40 percent, it's hollow and doesn't feel as sturdy as wood or composites. It bills itself as maintenance free, but it usually needs more cleaning than other decking materials due to its grooved surface. It is a good option for balusters and railings, however, and comes in a range of colors and styles.

### Saving Money on Your Deck

Save money on your deck by sticking to standard lumber sizes and avoiding odd shapes. You could also design your backyard living area to be part deck and part patio.

Opt for supported 2×4s for decking, because 2×6 boards are more expensive and tend to warp more frequently. Most people start to think about outdoor living spaces in the spring, and high season extends through the summer. If possible, book your job in the fall or early winter, depending on your climate; you should be able to get a better price.

### Patios

Patios, also called *lanais* in some areas, are concrete, tile, or stone-floored outdoor areas. They're great places to entertain or just relax with your family.

As with deck projects, spending some time planning will help you save money on your patio in the long run. Decide how you'll be using your patio. Will you have a table and chairs for dining? Will you have a barbecue, either built-in or portable? How else will your patio be used? Knowing your usage will help you plan the shape and size properly and avoid the dissatisfaction and waste of money.

Patio cost is less influenced by shape than decking is—the labor and materials don't change much regardless of shape. Size will affect cost, because additional labor and supplies will be needed. For sloped areas, you can also "step" your patio, using retaining walls to keep soil in place. This will usually require more excavating than is needed for decking, because you'll have to make the ground level in several areas. But it may be worthwhile if it will make your backyard more functional.

### Saving Money on Patios

Patios are generally constructed by marking out the area of the patio and digging out an area that's about 4 to 6 inches deep, depending on the materials used to create them, then laying the materials. Obviously, the easiest way to save money on patios is to choose less-expensive materials. Gravel is the least expensive, but it may not be as stable or pleasant to walk on. Stamped concrete is less expensive than pavers or brick, but it gives a similar effect.

Save on labor and materials by creating your patio yourself, if you're able, or book your contractors in the off months, as we discussed with decking.

## Landscaping Basics

Landscaping and gardening do more than make your neighbors envious. Well-planned plantings can make your house look great, help you save on utility bills, conserve water, and even give you food. You should plan your landscaping depending on your level of ability, goals, and, of course, your budget.

### Planning Your Landscaping

Planning your landscaping can be much like planning your house. The more time you spend at this stage, the more money you can save, and the more satisfied you'll be in the long run. Consider the following:

- **Climate.** Do you live in an area where landscaping is a year-round endeavor, or are you in a cold-weather climate where most plants, shrubs, and trees lie dormant in the winter?

- Exposure. Southern exposure gets the most sun, whereas northern exposure gets the least sun. Knowing which plants do well in various degrees of sunlight will help you make better selections.

- **Space.** The amount of space you have to plant will also dictate how you landscape your property. When you select plantings, be sure they're appropriate for the space. For instance, don't plant fast-growing trees, such as some evergreens, in a space they will quickly outgrow.

- **Soil.** Different plants do well in different types of soil, so take note of the types of landscaping your neighbors have, or read up on which plants are native to your area. The "locals" are likely to do better. You can also test your soil with inexpensive testing kits available at many garden centers. The results will give you more definitive information about soil makeup, which you can share with landscape professionals (or the free sources listed in the "Free Help" section later in this chapter) to give you solid information about which types of plants will do better in your environment.

- **Wish list/goals.** What do you want in your landscaping? Do you dream about eating vegetables from your own garden? Cutting roses from a backyard trellis? Plan for landscaping that will make you happy.

Also remember that landscaping evolves over time. As trees and shrubs get more mature, they add value to the property and create a beautiful, natural look that can't be duplicated overnight.

## Use a Professional or Do It Yourself?
As with choosing contractors to build your home, you have a variety of choices for your landscaping. If you have a green

thumb, you might want to try to plan it yourself. Or you could hire a landscape architect or landscape contractor to help you plan and plant your yard. Or much like hiring a building manager, you can hire a landscape designer or architect to create a plan and then install the landscaping yourself, saving the labor cost of planting.

### Free Help

Before you hire anyone to landscape your yard, take advantage of the free help that's available. You can find free help at the following places:

- **Flower and garden shows.** Throughout the country in the winter and spring, a crop of regional flower and garden shows pop up. Exhibits include professional landscape designers, architects, and contractors who plant elaborate gardens in the hope of landing business. Often their gardens and booths are staffed with professionals who are happy to answer questions—for free. So bring a pencil, paper, camera, and list of questions to record ideas and answers.

- **Garden centers.** Garden centers are great places to get ideas and free advice. Knowledgeable staffers can tell you which plants do well in what type of soil, exposure, and climate. Walk around and see which bushes, plants, and trees catch your eye. Ask questions and pick the brains of these professionals for ideas.

- **County cooperative extension offices.** One of the best-kept secret sources of free advice is the wealth of information available at county cooperative extension offices. These offices have access to agricultural and horticultural resources and have individuals and volunteers who can answer even tough questions about landscaping. Contact your county headquarters to locate your extension office.

- **Local garden clubs.** The much-maligned ladies and gentlemen of local garden clubs are also a terrific resource. Garden club volunteers are often very willing to assist new or wannabe gardeners with questions.

- **Local colleges.** Colleges and universities in your area may have students in landscape architecture, botany, or other related classes who could work with you to create a landscape and garden plan. Contact the main office and ask for the horticultural or agricultural education department.

**DON'T TRIP** on your **SHOESTRINGS**

Not all landscapers are created equal. The term *landscaper* can mean a guy with a lawn mower or it can mean a skilled professional. To be sure the person you consult is truly knowledgeable, look for titles such as *landscape architect* or *landscape designer*. Ask about schooling, and ask to see photos of past work. It's also a good idea to find examples of the work this landscaper has done in the area and speak to the individuals who hired him or her to find out how the landscaping has held up over the long haul.

## Landscaping that Saves You Money

Proper landscaping will make your environment more pleasant, but it can also save you money and improve the environment.

One way landscaping can deliver more than looks is if it delivers food for you and your family. Whether you love the taste of fresh-picked herbs, or you want to be able to feed your family from a backyard garden or fruit trees, your landscaping can usually accommodate your wishes. Fruit-bearing trees are often expensive and require regular maintenance, but planting annual vegetable gardens is a relatively easy proposition. However, be sure you have the time for maintenance and tending the plants for best results.

Your landscaping can actually help you save on your energy bills, too. How much? According to the U.S. Department of Energy, carefully positioned trees can save up to 25 percent of a typical household's energy for heating and cooling, and just three trees, properly placed around the house, can save an average household between $100 and $250 in heating and cooling energy costs annually.

Energy-efficient landscaping has three primary components:

- **Shade.** Shade from trees and large shrubs can help cool your home. According to the DOE, shading and evaporative cooling from trees can reduce the air temperature around your home. Studies conducted by the Lawrence Berkeley National Laboratory found summer daytime air temperatures to be 3° to 6°F cooler in tree-shaded neighborhoods than in treeless areas. The energy-conserving landscape strategies you should use for your home depend on the type of climate in which you live.

- **Wind reduction.** Using trees and shrubs to minimize wind against your home can save on energy bills. The most effective wind-reduction plantings are on the north and west sides of the home. Shrubs planted close to the home also help minimize snow accumulation next to the home, which can increase heating costs.

- **Water conservation.** The concept of using landscaping that requires very little water is called xeriscaping, which uses drought-resistant plants that don't need much water. In xeriscaping, traditional bluegrass or turf grass is replaced with hardier varieties of grass that don't need as much water, and mature trees and shrubs are preserved, because they generally need less supplemental watering. Mulch is used to retain moisture in the garden.

### Putting It All in Place

How do you put together this landscaping puzzle? You need to look at your plantings and start with the largest trees you have planned and work down in size from there. Consider the seasons in your area, and plan plant materials accordingly. In many places, it's possible to plan to have blooming plant material for most of the year, which can be a beautiful addition to any yard.

# Fence Me In

Depending on your municipality, if you have a pool in your backyard, you may be required to have a fence. Even if you're planning your fence to enhance décor, increase privacy, or provide additional wind protection, your local building codes will probably address some specifications your fence must meet.

Fencing ranges from simple chain link fences, which are about the least-expensive option, and usually least attractive, to extra-tall wood fences, to fences made from plastic or composite materials.

To reduce fencing costs, fence only the areas that need to be enclosed. Choose the least-expensive fencing material to serve your needs, and be sure that if you're using fencing to enclose an animal or to withstand harsh weather, it is durable enough to do so. Fencing is usually one of those projects that can be put off until later. So if your budget is tight, consider holding off on fencing your property, unless building codes require you to do so.

## Dollar-Saving Do's and Don'ts

- Save money by comparing costs of various materials and choosing those that are more economical.
- The longer your driveway, the more expensive your driveway and utilities will be.
- Team up with neighbors. Booking services in volume can save you money.
- Patios are generally less expensive than decks.
- Take advantage of free landscaping resources before paying a professional.
- Fence the smallest portion of yard that suits your needs.

# 12

# CONTRACTORS, ESTIMATES, AND SUPPLIERS

You've got a good game plan for your home. Now you need to line up the players who are going to make it happen. Although some home builders decide to take a total do-it-yourself approach, we advocate working with a variety of professionals, depending on your skill and the amount of time you have to spend on and complete the project.

## Understanding Contractors

Before you decide whom to hire, you should have a good understanding of the various kinds of contractors and what they do.

### General Contractor

A general contractor, or GC as they're commonly known, is basically a project manager. Using a general contractor means you'll have a one-stop shop for all, or at least most, of your building needs. The general contractor may have his or her own teams of workers, but more likely he or she will be hiring subcontractors to fulfill at least some or all the building requirements.

Some of the benefits of using a general contractor include the fact that he or she disburses payments to the various subcontractors. The GC will generally handle securing all your building permits (but not always, so be sure to check) and will oversee that the work is carried out to your specifications. To compensate for overseeing the project, the GC generally tacks on an up-charge of about 20 to 25 percent of the cost of the services and materials.

Another downside of hiring a general contractor is that you're usually committed to hiring the subcontractors he or she uses. Plus, if the contractor has too much on his plate, your project may suffer.

### Subcontractor

A subcontractor performs only part of your building project. The person or company often specializes in a particular area, such as masonry, excavating, plumbing, or electrical work. Subcontractors usually perform a very specific part of the project, for which you are often responsible for getting the permits and inspections. The benefit, obviously, is that there's usually no up-charge for their services, but they'll often tack on a percentage if they're responsible for buying materials related to your project.

The benefit of hiring subcontractors is that you can choose who works on your house. You can pick the best mason, best framing contractor, and best plumber your budget will allow. However, that's also the challenge—you're responsible for interviewing all these contractors and finding the best person for the job.

## Project Manager

Sometimes you can strike a deal with a general contractor or a subcontractor and modify his or her role to be a little more or a little less than it usually is. If you need the assistance of a project manager to advise you on the process of building your home, it may be a good idea to see if you can convince the general or subcontractor to act in that capacity.

A project manager can save you some headaches, but you may run into some of the same challenges of hiring a general contractor—a preference for certain subcontractors and scheduling issues.

## A Word About Contractor Licensing

According to the National Homebuilders Association, there are no hard-and-fast rules for the licensing of general contractors, and, in fact, many states don't offer general contractor licenses per se. In fact, according to www.contractors-license.org, a site recommended by the NHBA, 17 states require no license for general contractors. Most do have licensing requirements for plumbers, electricians, etc., as well as those contractors working with asbestos. Some states require registration rather than licensing, and others require certification or other credentials.

What does this mean to you? Well, the meanings are as varied as the requirements. Although some states have strict requirements for things such as continuing education and proof of insurance, others consider a contractor registered after he or she fills out a simple form and mails a check.

Licensing is a good sign, because it is likely that a contractor with many complaints or who has a history of deception would be refused a license. However, don't just take licensing at face value. Visit www.contractors-license.org for more information about the licensing requirements in your state.

### To General Contract or Not to General Contract?

Clearly, when you're building your home, you have many options you can turn to for assistance. Usually, the more time and expertise you have to devote to a project, the less contracting help you will need. Before you decide to take on the general contracting of your own home, ask yourself a few questions:

**Do you really have the time to do this?** General contracting your own home will become your full-time job for the next several months. If there's something to be done, you're the one who has to do it. When problems, delays, or mistakes happen, you're going to have to solve them. You need to visit your building site often and have enough flexibility that you can meet with your subcontractors and suppliers on a regular basis—sometimes at the drop of a hat.

**Is your job or schedule flexible enough to accommodate building your own home?** If you have a boss who checks your punch card and complains if you're a few minutes late, general contracting your own home could be a monumental challenge. You need to be flexible enough to meet with your contractors and inspect the job site while the workers are there. You'll need to field many phone calls and have the flexibility to deal with questions as they come up. If one of your contractors has a question about your framing and you can't get back to him until the next day, you've lost a day of work on your project. So be sure you're going to be able to give the home construction the attention it will most certainly need.

**Do you have any expertise in building?** It's far easier to act as your own general contractor if you have familiarity with the building process. If you don't know a trussed roof from a foundation footing, you'd better burn the midnight oil getting up to speed, or overseeing the building of your own home is going to be a challenge. Lack of expertise also means you're going to be more likely to make mistakes, so be aware of these challenges.

Are you comfortable handling negotiations and being direct with people? One of the primary ways you can save money throughout the homebuilding process is to be a good negotiator. But if you balk at asking for a better price or have a hard time pointing out problems when you see them, this isn't the best route for you to take. When you feel like the price is too high or the work isn't being done properly, you need to speak up. If that's a problem for you, hire someone to help you build your home.

Do you *want* to do this? Sure, it's great that you'll be saving as much as one quarter of your homebuilding expenses, but do you really want to be your own GC? Are you excited about the prospect, or does it fill you with a dull sense of dread? If you really don't have the desire to do this, don't take it on. It's a big challenge and difficult enough when you have a high level of enthusiasm.

Have we scared you off yet? No? Excellent! Just as you didn't settle for the housing options only available to you through pre-planned developments, you'll now have the opportunity to be sure the work is done to your exacting specifications and preferences. You'll have a great deal of satisfaction knowing you were involved in every detail of building your own home.

## On Your Own

Are there projects you want to take on yourself? Sure. You can do lots of things that will save you money and not put the construction of your home in jeopardy. Depending on your level of skill, you could try installing trim and molding, landscaping, painting and wallpapering, and installing lighting and fixtures.

There's really no limit to what you can take on yourself. It just depends on how able you are to manage the task. Just as you did when you thought about general contracting, think about whether you have the time, inclination, or ability to do

**DON'T TRIP** on your **SHOESTRINGS**

When you were pre-qualified for your loan, your lender should have made clear whether you are permitted to act as your own general contractor or whether you are required to hire a project manager or general contractor. Lenders like to feel as though their project is in experienced hands, so acting as your own GC may affect your financing package. If no mention has been made of this, double-check.

these things. It makes no sense to cut a few hundred dollars out of your budget to put up molding yourself if you're never going to get around to doing it. You've spent many times more on your house—why let unfinished business irk you every time you look at it? If you really aren't sure you'll do the work, consider hiring someone to do it for you.

## Finding Subcontractors

There's no shortage of contractors, so it should be easy for you to find professionals to interview. Here are a few places to start:

- **Friends and family.** As always, it's best to deal with someone whom you know firsthand has done good work.
- **Suppliers.** Often building supply companies have lists of approved or recommended contractors.
- **Chambers of Commerce.** Check with your local Chamber of Commerce to find many types of local businesses.
- **Local building associations.** The National Association of Homebuilders has local chapters throughout the country. Visit www.nahb.org or call 1-800-368-5242 to find an association near you.

If all else fails, you can always consult the ads in your local newspaper or the Yellow Pages. We've found subcontractors—and suppliers, for that matter—in the unlikeliest places. When Hank was visiting a friend and liked the brick work being done on the outside of the home, he walked over and got the contractor's number.

### Avoiding Contractor Horror Stories

Maybe you heard the story about a friend of a friend whose contractor demanded a big down payment on a project and then slipped off into the sunset with the money, never to be seen

**DON'T TRIP**
on your **SHOESTRINGS**

Even if your subcontractor comes with glowing recommendations, always get—and check—references. In addition, contact your local Better Business Bureau (find yours at www.bbb.org). If the contractor holds a license from the state, county, or municipality, she or he should post the license number. Call that licensing body, and ask if any complaints have been filed.

again. Gwen and Hank have good friends who hired a contractor to build an addition on their house. He opened the roof and then left their home exposed during the winter months. It took more than two years—and finally hiring a second contractor at additional cost—to complete the project.

It seems like everyone has a contractor horror story, but the truth is that the vast majority of contractors are honest business-people who try to serve their customers as best they can. Other-wise, they wouldn't be able to sustain a business.

To protect yourself from unscrupulous contractors, check references and licensing boards, as we've suggested. But also look out for these red flags:

- Inability to produce references or show similar jobs that have been completed in the past
- Demands for most or all of the job fee up front or demands for payment in cash
- High-pressure tactics, such as one-day-only pricing
- Indications that he or she skirts legalities, such as securing permits
- Inability to provide proof of insurance
- Reluctance to sign a contract
- Lack of familiarity with local building codes

Go with your gut feeling. If you feel as though the contractor isn't honest or won't do a good job, find someone else. It's im-portant that you feel confident in the person who is building any part of your home.

## Ask Good Questions

When you've narrowed down the field to a few good contrac-tors, interview them as you would anyone you intend to hire.

The contractors should be used to answering questions such as the following:

**Have you worked on projects like mine before?** It's helpful if a contractor has completed homes of a similar scale. If your contractor is used to more upscale homes than what you're building, he may not be as budget-conscious as you'd like. If she's only built smaller homes, she may not be prepared to take on your 4,000-square-foot dream home.

**Will you be the one doing the work on my property?** It's not unusual for a contractor to sell you on his experience and then have more junior employees or even subcontractors do the work. These people may not be as qualified as the contractor you hired. It's best if the contractor is hands-on, but at the very least, you want to be sure he supervises the work closely.

**Can you provide proof of workers' compensation and liability insurance?** Reputable contractors always carry this insurance, and you should never work with anyone who can't produce proof of insurance. If a worker gets hurt on the job or someone trips over a piece of plywood on your property and gets hurt, you could be sued and it could cost you thousands of dollars. We can't stress enough the importance of getting proof of insurance.

**Can you provide the names and contact information of five references and roughly when you did work for them?** We think five references will give you a good idea of the contractor's skill. Find out when the work was done and, if it was more than a couple years ago, ask for more recent references. Be sure you call the references and discuss what they liked and didn't like about the contractor. If possible, get their addresses and take a drive by their home to get a better feel for the quality of the work.

**Can you provide bank and supplier references?** This may seem to border on rude, but you want to be sure your contractor is financially stable and pays his bills on time. If he can't provide at least one banking and two supplier references, be wary.

**How big is your staff, and how many projects will you have booked at the same time as mine?** If the contractor is overbooked, you'll likely run into delays. Also ask what the contractor does when delays happen. If he says "We don't have delays," be skeptical. Delays can happen for a number of reasons that are beyond anyone's control.

**How can I contact you if problems arise?** You should have a way to reach your contractor quickly in case of an emergency. Many contractors will give you their cell phone number or a number to call after hours.

**Who is responsible for getting the permits for the work?** Sometimes the contractor will, but sometimes she'll expect you to do it. Even if your contractor does get them, ask whether the estimate or bid she submits will reflect the permit costs. Don't assume—ask.

**May I see a sample of your contract?** We go over the contract specifics later in this chapter, but getting a copy of the contract at the interview stage will give you time to go over it and suggest any additions or changes you want to make.

**How long have you been a contractor, and how did you learn your trade?** In some states, all a person has to do is hang a shingle and he can be a contractor. For some of the trades, such as electrical work and plumbing, the individual needs to meet certain training requirements. The longer the contractor has been in business, the better you should feel.

**How do you deal with changes in the project, if they arise?** Is the contractor willing to work with you to make your home exactly what you want it to be? A reluctance to deal with changes may signal a difficult working relationship.

**Can you provide me with a firm quotation and written timeline for the work?** Obviously, you want to be sure your contractor has a good understanding of what your project will cost to build and will commit to getting the work done in a reasonable amount of time.

**What kind of warranties do you offer?** Some contractors will provide various home warranties or guarantees for workmanship. You should get these in writing and consider the value of the warranty as part of the decision process of choosing your contractor.

This might seem like a lot of questions to ask, but you're much better off asking many questions up front and avoiding surprises later on. If the contractor seems impatient with answering the questions or doesn't answer them to your satisfaction, find someone else. There are plenty of contracting fish in the sea.

## Getting Bids and Estimates

When you have a few contractors you feel comfortable dealing with, the next step is to get estimates or quotations from them. An estimate is just that—an educated assessment of what the project is going to cost. A quotation, or quote, also called a bid, is a firm price that's usually attached as part of your contract. Your quote or bid will likely come in lower than the estimate, because contractors usually estimate a bit on the high side to be sure they cover themselves. The bid shouldn't be significantly different unless there's been a big change in the scope of work or materials specifications.

When comparing contractor pricing, you want to be sure you're comparing apples to apples. The best way to do that is to create a written outline of the project you submit to each of the contractors bidding on the job. That way, you know all the prices are based on the same scope of work and the contractor didn't misunderstand your request.

Ask for the estimate to be returned to you within a reasonable amount of time—1 to 2 weeks, depending on the complexity of the bid. Otherwise, you could be waiting for weeks to hear back from a busy contractor.

### Consider More than Price

Although we're very focused on saving money, sometimes going with the lowest-priced bid isn't the best way to go. If you find a contractor who has more skill or a better reputation at bringing jobs in on time and on budget, it may be worth the extra money to hire that contractor. Paying a bit more for a "sure thing" can be a wise decision in the long run. All things being equal, however, go for the lowest price!

Also, don't be surprised if you don't hear back from a contractor. When Hank and Gwen solicited roofing contractors for a job, Gwen set up six appointments for one day. Only three of the contractors showed up for the interviews, and only two of those submitted a bid. Sometimes contractors get busy or decide the project isn't a good fit for them. Always get bids from more contractors than you think you'll need.

It's also a good idea to request that the contractor offer any improvements or suggestions he or she think of. That way he or she feels welcome to make suggestions that could improve your project—and possibly save you more money!

Never sign a bid or estimate. Doing so could act as a binding agreement for service and could end up costing you money. The only document you should sign is your contract—after you've approved it.

## MONEY IN YOUR POCKET

When you negotiate with your contractors, ask for the option to buy your own materials. That gives you control over the brands you buy, and it can save you big bucks because you don't pay the contractor's markup on the materials. Ask the contractor to provide you with a detailed list of the materials needed at the same time he or she provides the estimate. Purchasing such high quantities may qualify you for wholesale discounts at various building supply companies. Shop around, and get the best deals on the best brands. Also ask your contractor for tips, product preferences, or suppliers. He may be able to steer you in a good direction. A standard contractor's discount is anywhere from 5 to 15 percent. Average that—10 percent—on a $40,000 materials order, and you've saved $4,000.

You may also consider purchasing the materials on credit cards that offer reward programs. Because you'll be spending so much money, you may earn enough points to qualify for gift certificates, appliances, or airline tickets. Just be sure to pay the balances right away so you don't incur finance charges.

**Savings for You: $4,000**          **Running Total: $76,930**

### Ch-Ch-Changes

Sometime, somewhere in your construction, you're going to want to make a change. It may be that a room is a bit smaller than you like or that you'd prefer to add or enlarge a closet.

Of course, the more changes you have, the more money will be added to your cost, but you and your contractor should come to an agreement about how changes will be handled and at what cost. Changes made before construction starts should cost you less than if you change your mind after the work has been done. Also, to avoid any misunderstanding, submit all changes in writing with clear specifications.

### Getting a Better Price

As we said earlier, if you're not good at being direct about what you need and asking for a better price, you may have trouble with the process of overseeing your home's construction. It's important to remember that sometimes the easiest way to save money is to simply ask.

Here are some tips on how you can save money with your contractors.

- **Be frank about your budget.** Give your contractor a ballpark budget so he knows whether to suggest elaborate changes or simply do the bare-bones work. However, don't give him your exact budget. Always reduce the number by a 10 to 20 percent reserve in case there are overages.

- **Ask for a match.** If you receive two bids but would prefer to work with the contractor who submitted the higher bid, don't be afraid to ask the contractor to match the price of the lower bid. Depending on whether he needs the work or not, he may be willing to negotiate on price.

- **Limit changes.** It may seem obvious, but the more changes you make, the more money it's going to cost. However, as the project starts to come together, you're going to be tempted to make those changes. Before you jump the gun, ask yourself if the change is something that's really worth the money to modify. If it is, go for it. If it's a whim, let it pass.

- **Schedule off-season work.** In different areas of the country, contractors have different seasons in which they work. For instance, where Gwen and Hank live in New Jersey, we were able to schedule some contractors for fall, after the busy summer season started to slow down. As a result, we were able to negotiate slightly better rates from the contractor, whose demand had fallen off as the weather cooled.

  Be prudent about scheduling off-season work, though. It's probably not a good idea to start building a home in Minnesota during the winter or in the South during hurricane season.

- **Allow plenty of time.** Building a home takes time—lots of time. In fact, it's almost always going to take more time than you think it will. Be sure to allow yourself time for unexpected delays due to weather, materials issues, inspection problems, and other factors. (We cover this in more detail in the next chapter.) If you make your schedule too tight, you may incur rush charges and extra costs for rescheduling your contractors.

**DON'T TRIP** on your **SHOESTRINGS**

Part of the negotiation process includes determining how much will be paid at various intervals of the job. Never pay more than a small percentage up front. In fact, it's a good idea to hold off paying your deposit until the first day the contractor shows up to work—incentive!

Beware of contractors who want money for materials up front. Most reputable contractors have lines of credit with their suppliers. So if a contractor needs you to float his expenses, it's likely he doesn't have a good financial record. Be wary.

## Lien, Too

An important part of your agreement with your contractor will concern liens against your property. It may seem unfair, but if a contractor hires a subcontractor or orders materials from a supplier for your project and doesn't pay them, these subcontractors and suppliers can place a lien against your property—even if you already paid the contractor the money for these materials and labor!

A lien is the right to take possession of a piece of property to fulfill or secure a debt. That means the subcontractor hired to pour your foundation holds a claim on your property until his debt is fulfilled. Liens can affect your lending agreement, because an entity other than your primary lender holds a claim to your property. In addition, liens may damage your credit rating.

Of course, you can always take the contractor to court and try to sue for your money back, but that's a lengthy and expensive proposition. You'll save yourself a world of headaches, delays, and money if you require that all contractors disclose the subcontractors and suppliers they use, provide you with their contact information, and agree to secure a lien waiver—a document that states that the vendor has been paid and relinquishes all rights to sue you or try to hold you responsible for payment—from each subcontractor.

## Permits and Inspections

Another area you need to reach an agreement with your contractors on is who will secure building permits and schedule inspections. Both of these are primarily interactions with your municipality's building department.

Permits, in the simplest terms, are documents that state that your municipality has granted you permission to proceed with your project as per the specifications you have submitted. Permit fees are generally a few hundred dollars and can increase depending on the project.

---

### MONEY IN YOUR POCKET

Check to see if you can bundle many permits into one or two. Instead of paying $300 for your framing, roofing, and foundation permits, check to see if you can bundle them all into one permit. So if you include your deck and your pool on your home's building permit, instead of applying for them separately, you'll save the additional fees for separate permits. If the fees for the above are $300 each, you've saved $600.

**Savings for You: $600**          **Running Total: $77,530**

## Signing the Contract

You've gotten your bids, have decided on the best price, and feel comfortable with the contractor doing the work on your home. Now it's time to put it all in writing.

We can't emphasize enough how important it is to get all the specifications, agreements, and prices written into the contract. (See Appendix C for a sample contract.) This document will be the reference to settle any disputes that may arise, and the more detail it includes, the less room there will be for misunderstanding. Your contract should include the following:

- Your name and the address of the property on which the work will be done.
- A precise description of the work covered in the contract.
- A list of materials, including as many identifying details, such as color, grade, size, brand, product numbers, etc., as possible.
- Date on which the project will start.
- Date on which the project will be completed and a schedule of when the work will be done. The latter will help you determine if the project is on track to be finished on time.
- Assignment of responsibility for securing building permits and inspections.
- A provision that the work will be done to meet local codes and is guaranteed to pass inspections. If not, you should have a provision that the work will be corrected to pass inspection.
- A detailed list of the costs of all labor, materials, and fees.
- A schedule of payments, including down payment and amounts to be paid at specific completion points. If your lender is disbursing payments, that should be outlined in the contract, and the contract should state that payments

will not be made until the work passes any required inspections.

- An explanation of how changes will be handled—both preconstruction and after installation—and the rate at which charges, if any, will be incurred. You should have a provision that changes need to be submitted in writing to avoid increased charges based on verbal misunderstandings.

- Verification of who will be doing the work and that the contractor will provide you with all names, addresses, and contact information of any subcontractors and suppliers used. Contractors should also be responsible for securing lien waivers from any of their subcontractors and suppliers before payment is disbursed.

- A cleanup clause. The job site should be kept in reasonable order to avoid injuries, and the contractor should be responsible for carting off any debris left over from the project.

- Standard indemnification clauses that hold you harmless from action taken against the contractor for injuries or other issues.

- Remedies for lack of performance. You should have a way to break the contract if your contractor isn't holding up his end of the deal. A provision that the contractor remedy problems within a specific amount of time or allow you to break the contract with no further payment can save you problems in the long run, and acts as motivation for the contractor to abide by the agreement.

You can also request a delay clause. That means that if the contractor doesn't complete the work within the specified time frame, you get a portion of the fee rebated back to you, depending on the length of the delay. For instance, you could charge the contractor $50 per day for every day the project is delayed. Of course, this wouldn't apply to delays beyond the contractor's control, such as weather, strikes, acts of God, and the like, but it is an incentive for an overbooked contractor to make your project his priority.

Part of the contract package should be a certificate of insurance for the contractor's workers' compensation and liability policies. In addition, any verbal agreements you have with the contractor should be recorded in the final contract. (See Appendix C for sample general contractor and subcontractor agreements.)

Finally, it's always a good idea to have your attorney review your contract to protect your interests. It probably won't cost you more than $100 to $200, but this review could save you thousands of dollars if the agreement properly protects you from something that goes wrong.

## Money Money Money

It's important to be fair to your contractors and pay them in a timely fashion, as per your agreement. Depending on the project or the contractor, you can expect to pay between 5 and 20 percent of the contract fee up front. (Beware of contractors who ask for more than that.)

After that, your contract should specify specific periods at which the contractor will be paid. For instance, your framing contractor may be paid as follows:

15 percent at the start of the project

25 percent when the first floor is framed

25 percent when the second floor is framed

20 percent when the sheathing is completed

15 percent when the job is done and all inspections have been completed and passed

**DON'T TRIP**
on your **SHOESTRINGS**

Your contractor may have subcontractors, and if they're not paid they could place a lien against your property, even if you've already paid the primary contractor. Stipulate in your contract that the contractor will hold you harmless from all liens against the property and that he will secure lien waivers from any subcontractors used on the project.

Keep your payment schedule consistent with the amount of work to be done. For example, your contractor should be scheduled to receive about 45 percent of the fee when half of the work is done. If you pay more than that, you won't have the funds available to hire another contractor to finish the job if something goes awry.

Always hold a reserve of about 10 to 20 percent of the entire project fee to be paid after the job is done. Allow time for all inspections to be completed, and be sure you have all your lien waivers before you fork over the last of the cash.

## Good Contractor Relations

Fostering good relationships with your contractors will save everyone headaches down the line. Plus, good contractor relations make it less likely that you'll have problems that could cost you money. Here are some hints to getting along with your contractor:

- **Know your stuff.** Brush up on the techniques being used so your contractor has to do a minimum of hand-holding. Read books or websites about framing, foundations, and other building components so you understand your options and can recognize and recommend money-saving solutions and ideas.

- **Keep communication open.** Speak with your contractor regularly. Ask questions, and listen to his or her concerns.

- **Be accessible.** If your contractor is on-site and has a question, being inaccessible for a day could cause a significant delay in your project.

- **Be visible.** Visiting your job site often lets your contractor know you're serious about monitoring the quality and progress of the work. Absentee owner-builders may find that their contractors don't show up as regularly or pay as close attention to detail as those of owner-builders who are frequently checking in.

- **Be reasonable.** Pitching a ranting, raving fit when something doesn't go your way isn't productive for anyone. If problems arise, deal with them calmly and rationally.

- **Pay promptly.** Contractors and suppliers are used to waiting for payment, but that doesn't mean they should have to wait longer than necessary. Be sure your contractors are paid as per the agreement you've made. Plus, delays in payment could end up costing you late charges.

### When It's Not Working Out

Building your home with a contractor is a little like a marriage. You're likely to have times where you don't see eye to eye. That will happen, and that's why it's important to keep your communication with the contractor open and to speak up when things aren't to your liking.

Sometimes, however, in spite of your best-laid plans, the contractor you chose so carefully won't work out. Perhaps he doesn't listen to your concerns or his work is substandard. Some common complaints about contractor performance include the following:

- Contractors not reporting to the job site when expected
- Sloppiness, either in the work or in keeping the job site properly maintained
- Incorrect work or failure to adhere to plan specifications
- Lack of accessibility when you have questions or concerns
- Delays due to overscheduling or contractor mistakes
- Failure to adhere to contract terms

Whatever the problem, your best bet is to try to work it out. Sometimes it's not easy to tell your contractor that you're dissatisfied with his work, but it must be done to ensure that your home is built within the time frame, budget, and specifications upon which you've agreed.

To protect yourself, act quickly and follow the steps outlined in the following sections.

### Talk It Out

Have a calm but straightforward conversation with your contractor about your concerns. Try to come to a successful resolution for both of you.

### Write It Down

After you've had a conversation with your contractor, document the conversation and its result in a letter, whether you were able to come to an agreement or not. Send the letter, including the agreement you made or the remedy you demand within a specific time frame, to the contractor along with some sort of proof of receipt. Certified or overnight mail will both provide you with proof that the contractor received the document.

Sending the confirmation in writing shows you're serious about getting it corrected. It also protects you, showing that you made a good-faith effort to rectify the situation, should you need to seek external remedies such as court action.

### Exercise Your Rights

If you can't come to an agreement and the contractor won't remedy the situation, you may need to exercise your contract's "out" clause. This will usually require notification in writing. Again, send the notification via certified mail or overnight to provide you with proof of receipt.

If you've paid the contractor for more of the work or materials than he's completed or supplied, you may need to take legal action to recoup your cash. (This is why it's a good idea to pace your payments with the progress of the project.)

If the amount you're due falls below a certain amount, you can take the matter to your state's small claims court. Small claims courts will make awards ranging from $2,500 up to $10,000, depending on your state, and you don't need an attorney to file. If the amount you're due is more than your state's small claims court allowance, you may want to consult an attorney about your options. In any case, be sure you bring all documentation related to the project to prove your case.

## Finding and Working With Suppliers

Although much attention is paid to finding the right contractor, it's also critical that you find the right supplier, if you'll be buying materials yourself—and saving that contractor's markup.

Your supplier is the company from which you'll be buying the materials to build your home. You may choose one multipurpose supplier, or you could also choose several smaller or specialty suppliers.

### Material Things

If you've received a *take-off list*—the list of supplies and materials you need to build your house—from your architect or the company that drew your plans, that's a good place to start. As with contractors, you'll want to get a few bids on supplies. It's very likely that you'll find that one supply house has great prices on one item, such as lumber, but is very expensive when it comes to another, such as doors. Don't feel compelled to purchase all supplies from one supply house. Shop around, and purchase from the suppliers who give you the best prices and terms.

If you don't have a take-off list, it's likely that the supply house will do one for free as a service to try to get your business. It's best to work from one take-off list to be sure you're comparing apples to apples on the bids. If one supply house estimates that you'll need 200 sheets of plywood and another estimates that you'll need 250, that could mean a significant difference in both the unit and bottom-line pricing. So get one take-off list and have subsequent suppliers bid on it. You should also ask them if they see any discrepancies between the take-off list and what they would estimate for the job. This will give you a set of checks and balances. Of course, it's a good idea to run the take-off list by your contractor or contractors before you place the order to ensure that they don't see mistakes or have additional recommendations to make. Also remember the difference between estimates and quotes—be sure you're dealing with quotes when you consider pricing.

### Supply House Questions

Price is important, but you also need to consider some delivery terms. If your supplier offers you a great price but charges an exorbitant delivery charge every time materials need to get to the site, you could easily wipe out any savings on materials costs. Some of the questions you'll want to ask your supplier include the following:

**How long does it take for materials to be delivered?** Get an accurate estimate of how long each type of material takes from ordering to delivery. You'll want to be sure materials such as lumber and wallboard can be delivered quickly in case your contractor runs low. If lumber can't be delivered within a day or two at the most, find another supplier.

**What is your return policy?** Because virtually no take-off list is 100 percent accurate, you should be able to return some materials, such as excess lumber and plywood, that are uncut and in new condition. At nearly $20 per sheet, returning excess plywood can quickly add up to big savings. Find out which materials are returnable and which aren't. Order those materials that aren't returnable very conservatively, because you can always order more.

**Do you offer warranties on materials?** Find out if the supply house makes any warranties on materials, or if such assurances come from the manufacturer. Also find out what the company's policy is on replacing damaged or substandard products. For instance, if you receive a batch of lumber that is weathered and moldy, you're going to want to be sure you can exchange that for new lumber.

**Is there a delivery charge?** Some supply houses charge for delivery, whereas others have free delivery for orders over a certain dollar amount. Find out the specifics of delivery charges, and then make decisions about how your materials will be delivered. It's best to have them delivered as needed, because then your materials aren't exposed to the elements. This also reduces the chances of materials theft. However, if your supplier charges stiff delivery charges, plan your deliveries so you maximize your savings. You can then use a construction fence and lock to keep the site somewhat secure and cover the materials with plastic sheeting to keep them pristine. (Keep in mind, however, that the fence and sheathing could cost you as much as $100 or more, so that should be factored into your decision.)

**What kind of and how many materials are in stock?** Different supply houses keep different materials in stock. In-stock materials can usually be delivered quickly, whereas special orders may take several weeks. If you're going to be ordering large amounts of something, be sure the house will have it on-hand.

**Is that your best price?** Negotiating with supply houses isn't much different from negotiating with contractors. You may be able to save additional money just by asking. Or if you find a supply house that you like but have a better price on an item from another supply house, ask the preferred house if they'll match the price. They may do it to keep your business.

Owner-builders usually qualify for builder's pricing, which is less than retail pricing. You'll often do better on price by working with a dedicated sales rep at a supply house. This person is familiar with your project and may have more leverage to get you better pricing, because he or she knows your history and project.

## MONEY IN YOUR POCKET

Many supply houses will extend credit to you and allow you to pay for your supplies according to a predetermined schedule. The account may be set up like a credit card, where you pay a certain amount of money every month, or it may be payable within a set amount of time, such as 120 days. You'll pay interest on the balance, which will add to your cost, but being able to float that expense until your next draw may make that extra cost worthwhile.

If you have the cash, you may be able to get a cash discount by paying for your supplies up front. Sometimes supply houses will knock as much as 3 percent off your bill if you pay as you order. This can add up to significant savings—as much as $1,050 on $35,000 worth of supplies.

**Savings for You: $1,050**          **Running Total: $78,580**

### Delivery Basics

After you've selected your supplier and negotiated your prices and payment, you still need to stay on top of the process. Be on hand when supplies are delivered to ensure they are of acceptable quality and the proper amount of supplies have been delivered. Also check your bill to be sure the billing matches the amount of materials delivered and the billed price matches the quotation. Mistakes happen, and you don't want to inadvertently pay more than necessary.

**DON'T TRIP**
on your **SHOESTRINGS**

Material theft can be a big problem—and a big expense. If you can, only have the materials you need delivered to the site, and be sure the site is surrounded by a construction fence (required by many municipalities, anyway) with a lock. If you have nearby neighbors who seem trustworthy, ask them to keep an eye on the lot. After your house is secured by sheathing, doors, and windows, it may be a good idea to move materials inside to keep them safe.

Having to buy more materials because of theft can add a big expense to your project, so be sure you secure your materials as well as possible.

## Dollar-Saving Do's and Don'ts

- Realistically evaluate whether you have the time and flexibility to be your own general contractor.
- Issue bid requests so you're comparing estimates for exactly the same labor and materials.
- Buy your own materials to avoid a contractor's markup.
- Don't be afraid to negotiate and ask for a better price.
- Get all agreements and warranties in writing to protect yourself now and later.
- Ask for builder's discounts, credit, and prepayment discounts from your suppliers.
- Secure your site to ensure you don't have to repurchase stolen materials.

# 13

# ESTIMATING, SCHEDULING, AND BUDGETING YOUR PROJECT

All the planning in the world is useless if you can't afford the final result. It's important to compile your bids and create a detailed estimate of the project, from clearing the first tree to installing the last appliance.

Estimating your project is a big job and will require making some of the tough decisions we've discussed in other chapters. But you can make it a bit easier and ensure that both your budget and your timeline remain intact.

## Get Organized

It's nearly impossible to stick to a schedule or keep track of all your bids if you don't take some time to get organized. Having all your notes, photos, documents, contracts, estimates, bids, and other paperwork handy can save you lots of time—and the massive headaches caused by lost documentation. Here are some approaches that have worked for us:

- **File cabinet drawers.** Set up a specific drawer to house all the documentation related to your project. Keep separate vertical files by various subcontractor or by the phase of the work (foundation, framing, siding, etc.).

- **Binders.** Keep your documents in a series of three-ring binders. Use a three-hole punch to ensure that all documents can easily be added to the binders. You can have one large binder, or separate the documents into smaller binders by job or project phase.

- **File boxes.** If you prefer to keep your files portable, invest in some file boxes at your local office-supply store. Use these like you would use a file cabinet drawer, keeping vertical files of various jobs or contractors.

- **Computer files.** Love using the latest technology? You can always scan your documents into files on your computer. This paperless approach enables you to organize your information, which can then be transferred to a CD or disk. If you choose this option, however, *be sure* to back up your information regularly, or one computer virus or malfunction could erase months of careful organizing.

Whatever you do, find a system that works for you. If a question arises later in the project, you need to be able to produce documentation to prove your point. If you don't have it, you just might be out of luck.

## Get Ready for Take-Off

Your take-off list is the list of materials you're going to need to build your house. Compiling a take-off list requires that you have your plan finalized—any changes are likely to affect the list and affect your cost.

It is possible to compile your take-off list yourself, but we don't recommend it unless you have a great deal of familiarity with the construction process. If you know the difference between a J-bolt and a joist or can spot a plumb wall from 50 feet, go ahead and work on your own list. You can, however, save yourself a great deal of time by taking advantage of some free assistance in compiling your list:

- **Your contractor.** If you're using a general contractor, part of his job will be to create the take-off list. Have him share it with you, though, and be sure it's as specific as possible, including brands, sizes, model numbers, and the like. When you have the list in hand, you may be able to find ways to cut costs, such as opting for a less-expensive roofing shingle or changing the type of flooring you're going to have installed.

- **Building suppliers.** Supply houses will often take your plans and create your take-off list for free as a part of their estimate. It's part of their selling process; they hope you'll then order the materials through them. It's a good idea to have at least two or three suppliers run your list and compare them. Even experts make mistakes, and having a few take-off lists makes it easier to spot anything one of the bidders may have missed. Also be sure to run your list by the various contractors who will be using the materials to ensure they agree with the quantity and quality of the products. If your contractor can't or won't work with the materials you've chosen, you'll just be adding delays.

- **Computer programs.** Some home-planning software packages have an option to create a take-off list. This can be a good jumping-off point, but it's also a good idea to run the list by your contractor or supplier to ensure that it's accurate. In addition, the program probably won't indicate brands or colors, which are critical to specify when you place your orders with suppliers.

Of course, if you purchased stock plans, you can also purchase your take-off list from the company that supplied them, but these are notoriously inaccurate. An architect can also generate a take-off list, but why pay for such a list when you can get it for free from your contractor or supplier?

## MONEY IN YOUR POCKET

If your contractor supplies the materials for the job, he or she will often tack on a 20 percent up-charge for doing so. This covers the time it takes to place the order, compensates the contractor for carrying the debt, and gives the contractor a profit on the materials provided. By ordering the materials and carrying the cost yourself, you could save that 20 percent. On $35,000 worth of materials, that's a whopping $7,000.

**Savings for You: $7,000**          **Running Total: $85,580**

## What's It Worth?

As you estimate your project, keep in mind what the home is going to be worth after you've finished it. Your lender will want to know that your finished project will have at least some equity in it and that its value will be in keeping with other homes in the area. This protects the lender in case the loan goes into default.

Local real estate agents, your lender, and even your tax assessor can help you find comparable property sales in the area. Compare your home to recent nearby sales of homes of a similar style, square footage, and features to get an idea of what your home will be worth. If your home is going way over budget and will cost more to build than it's worth, you may run into trouble with your lender, who may want to see you invest more of your own money into the construction. Avoid this possible pitfall by watching your budget closely and being sure your home value will not be significantly more or less than similar homes in the area.

## Look Out for Hidden Costs

As you assemble your final cost estimate, look out for hidden charges. Labor and materials costs are somewhat obvious, but some costs are often overlooked, such as inspections and permits.

Often, you'll need to pay for the inspector to come out and approve the work you've done. Check with your town's building department for a schedule of these inspection costs. You may also have inspection fees from your lender, who will inspect the property before releasing each disbursement of funds.

Permit fees are often overlooked in the budgeting process, but they can add up to a few thousand dollars, depending on the area and the work to be done.

Find out from your municipality and lender about any other hidden charges you'll incur along the way.

## Keeping a Reserve

You should build a 10 to 20 percent reserve into your budget that you keep "just in case." If you have very detailed specifications and take-off lists that include details down to the brand, size, and SKU number (product identification number) of each material, and if you feel very comfortable with your contractors, you may be fine with a 10 percent reserve. If you have less-detailed specs, opt for keeping a 20 percent reserve.

## GO FIGURE

You'll want to create an estimate sheet to give you an at-a-glance format in which to see your costs estimated. Here is a sample format, which you can also easily set up as a spreadsheet. Doing so will let you easily see how changing some estimates (by reducing the scope of work or using a less-expensive material) will affect your budget.

| | Permit | Materials Supplied by Contractor | Materials Self-Supplied | Labor | Inspections | Other Costs |
|---|---|---|---|---|---|---|
| Planning | $ | $ | $ | $ | $ | $ |
| Land | $ | $ | $ | $ | $ | $ |
| Excavating | $ | $ | $ | $ | $ | $ |
| Foundation | $ | $ | $ | $ | $ | $ |
| Septic system (if any) | $ | $ | $ | $ | $ | $ |
| Well (if any) | $ | $ | $ | $ | $ | $ |
| Framing | $ | $ | $ | $ | $ | $ |
| Roofing, gutters, and leaders | $ | $ | $ | $ | $ | $ |
| Windows and doors | $ | $ | $ | $ | $ | $ |
| Electrical | $ | $ | $ | $ | $ | $ |
| Plumbing (including bathroom fixtures) | $ | $ | $ | $ | $ | $ |
| Heating and cooling | $ | $ | $ | $ | $ | $ |
| Siding | $ | $ | $ | $ | $ | $ |
| Insulation | $ | $ | $ | $ | $ | $ |
| Wall board | $ | $ | $ | $ | $ | $ |

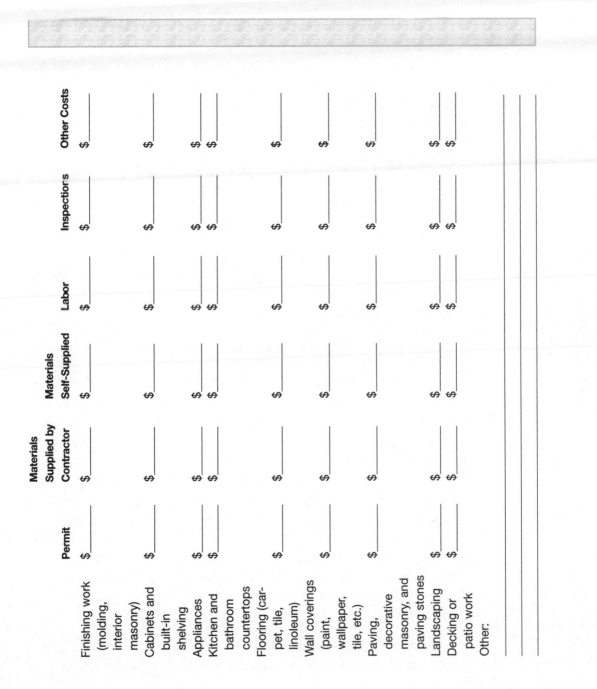

| | Permit | Materials Supplied by Contractor | Materials Self-Supplied | Labor | Inspections | Other Costs |
|---|---|---|---|---|---|---|
| Finishing work (molding, interior masonry) | $ | $ | $ | $ | $ | $ |
| Cabinets and built-in shelving | $ | $ | $ | $ | $ | $ |
| Appliances | $ | $ | $ | $ | $ | $ |
| Kitchen and bathroom countertops | $ | $ | $ | $ | $ | $ |
| Flooring (carpet, tile, linoleum) | $ | $ | $ | $ | $ | $ |
| Wall coverings (paint, wallpaper, tile, etc.) | $ | $ | $ | $ | $ | $ |
| Paving, decorative masonry, and paving stones | $ | $ | $ | $ | $ | $ |
| Landscaping | $ | $ | $ | $ | $ | $ |
| Decking or patio work | $ | $ | $ | $ | $ | $ |
| Other: | | | | | | |

Don't count this reserve toward paying for any phase of the project. Rather, put it aside and forget it's there until one of your budget items runs over. If you get through a significant portion of the project and find that you don't need the reserve, then celebrate! You can use the extra money for a variety of happy reasons, including the following:

- Upgrading appliances or countertops
- Upgrading exterior features or landscaping
- Installing a swimming pool or spa
- Paying down the principal on your loan

Or you can save it for upgrading interior features later on. If you put the money aside, just be sure it's not used for frivolous reasons. Be sure it works as an investment toward improving the equity in your home.

## How Long Will It (Really) Take?

One thing's certain whether you're building a home with a general contractor or overseeing the project yourself: however long you think it will take, it will take longer. We've heard it all before: you've made your schedule and checked it twice. You have great contractors. There's no way anything could go wrong, right?

Wrong. There are myriad reasons why a project will take longer than scheduled, including weather, contractor overbooking, missing or wrong materials or supplies, unanswered questions or homeowner inaccessibility, and prerequisite work delays (painting contractor delayed because drywall contractor hasn't finished his work). Those are just a few. The good news is that there are many ways to short-circuit these delays, including the following:

- **Pad your schedule.** Don't book your contractors back to back. Instead, allow a few days to a week between when you expect one contractor to finish and the next to begin. The bigger the job, the more time you should pad. For instance, you might want to add a 2- to 3-week allowance for framing, which is one of the biggest jobs in the construction process and also susceptible to weather issues.

- **Order materials in advance.** One of contractors' pet peeves is when they agree to let an owner-builder order the materials directly and then the materials are not available when it's time to work on the job. If you want imported tile for your bathroom or a special-order appliance for your kitchen, order it in plenty of time for it to arrive. You're far better off renting a storage facility to house the material for a couple months, at a cost of a couple hundred dollars, than you are holding up a contractor, which will have a domino effect that could cost you thousands. Be sure your materials are secure so they won't be stolen or vandalized, which could also cause delays in your project and force you to have to reorder your materials.

- **Plan for weather.** Let's face it: It's not going to be bright and sunny every day of your building process. If you're in the Southeast, you may have to deal with hurricanes. If you're in the Midwest, bitter cold and snowy winters may not be the best time to build. Be familiar with the weather patterns during the time of year you plan to build, and adjust your schedule to allow for rainy seasons, snow and ice, and other weather-related hold-ups.

- **Be specific.** If you want extra lighting in your home office, say so. If you're going to make one room a media center that needs extra wiring, speak up. If you want a specific type of roofing shingle, be sure your contractor knows. In other words, the more specific you can be, the better the chance the job will be done to your specifications the first time around. Do-overs and additions add time and cost.

- **Be a squeaky wheel.** About a week to 10 days beforehand, check in with your contractor to be sure he or she is still on schedule for your job. Ask your contractor to come out and check the job site beforehand to be sure everything's in order and that he or she has everything needed to avoid delays. Check in a few days ahead of time, then again the day before. Don't be a pest—keep the check-ins brief; but the squeaky wheel gets the contractor to come out on time. The same tactics can be used with suppliers to be sure delivery deadlines are met.

- **Allow time for inspections.** If inspection departments get busy, it becomes your problem because your job can't continue until what's finished has passed inspection. Get to know the inspectors in your town's building department, and contact the department to ask them how much time they need to schedule an inspection so you know how much time to build into your schedule.

- **Pay promptly.** Contractors are more likely to give priority to customers who pay their bills promptly. If you are able to, guarantee your contractor payment within 24 hours of passing inspection. Quick money may mean quicker service.

- **Schedule work during slow times.** If you're working with a flexible schedule, check with your contractors to see if it makes sense to book the work during slow periods. If a contractor's business falls off in the autumn months and you can wait, you might find yourself less likely to meet with overscheduling delays if you book the contractor during his or her slower time. However, discuss any weather issues with your contractor—you don't want to trade overscheduling delays for weather delays!

- **Ask your contractor(s) for ideas to avoid delays.** The best people to ask about avoiding delays are the people doing the work. They can give you tips for keeping various parts of the job on schedule. For instance, the framing contractor may be able to build a "chase," or an unobstructed channel that acts as a conduit for plumbing and electrical work. This could prevent delays in installing drywall if your plumber or electrician takes longer than expected.

## GO FIGURE

Whether you're working with a general contractor or are self-contracting, it's important to have a schedule of the phases of construction and an estimate of how long each will take. If you're working with a GC, you need his or her input to create the schedule. If you're self-contracting, much depends on the availability of the subcontractors you use.

| | Number of Days | Dates |
|---|---|---|
| Planning | _____ | _____ |
| Creating/finalizing plans | _____ | _____ |
| Land | _____ | _____ |
| Excavating | _____ | _____ |
| Foundation | _____ | _____ |
| Septic system (if any) | _____ | _____ |
| Well (if any) | _____ | _____ |
| Framing | _____ | _____ |
| Roofing, gutters, and leaders | _____ | _____ |
| Windows and doors | _____ | _____ |
| Electrical | _____ | _____ |
| Plumbing (including bathroom fixtures) | _____ | _____ |
| Heating and cooling | _____ | _____ |
| Siding | _____ | _____ |
| Insulation | _____ | _____ |
| Wall board | _____ | _____ |
| Finishing work (molding, interior masonry) | _____ | _____ |
| Cabinets and built-in shelving | _____ | _____ |
| Appliances | _____ | _____ |
| Kitchen and bathroom countertops | _____ | _____ |
| Flooring (carpet, tile, linoleum) | _____ | _____ |
| Wall coverings (paint, wallpaper, tile, etc.) | _____ | _____ |

Paving, decorative masonry, and paving _____  _____
  stones
Landscaping                    _____  _____
Decking or patio work      _____  _____
Other:

_____

_____

_____

## Overlapping Jobs

Of course, some jobs can be completed concurrently, saving time in your schedule. As you plan the work and schedule your subcontractors, you might take an approach like the following, which allows ample time allotments:

**April:**

- Finalize house plans and submit for municipal approval.
- Determine necessary permits.
- Collect bids from excavating, foundation, and framing contractors.
- Choose bids and begin scheduling contractors.

**May:**

- Clear the site.
- Mark the foundation perimeter.
- Dig the foundation.
- Strip the topsoil.
- Pour the footings.
- Begin constructing the foundation.
- Order the foundation and framing materials.

- Order the windows and exterior doors.
- Schedule any necessary inspections.

### June:

- Finish the foundation construction.
- Begin framing the walls of the basement (if applicable) and first floor.
- Order plumbing, electrical, and heating and cooling supplies, including bathroom fixtures.
- Schedule any necessary inspections.

### July:

- Frame the second floor and roof.
- Sheath the exterior of the house.
- Begin the electrical and plumbing work.
- Order any special fixtures, materials, cabinets, or appliances that need long lead times.
- Order the siding.
- Schedule any necessary inspections.

### August:

- Affix the roofing materials.
- Install the windows and exterior doors.
- Finish the plumbing and electrical work.
- Begin installing the heating and cooling systems.
- Order the insulation.
- Schedule any necessary inspections.

**September:**

- Install the siding and exterior trim.
- Install the gutters and downspouts.
- Finish the heating and cooling systems.
- Schedule any necessary inspections.

**October:**

- Insulate the walls.
- Begin the drywall.

**November:**

- Finish the drywall.
- Begin the molding work.
- Install the cabinets and counters.
- Begin the flooring work.
- Begin the tiling work.

**December:**

- Finish the molding.
- Finish the tiling work.
- Install the bathroom fixtures and appliances.
- Begin priming, painting, and installing wall coverings.
- Schedule any necessary inspections.

**January:**

- Continue the finishing work, such as electrical outlet covers, etc.
- Install the flooring.
- Schedule any necessary inspections.

**February:**

- Move in!

Of course, your schedule will vary and may take less time or more time than this example. The size and complexity of your home, the time of year, the weather, and the ability of your general contractor or subcontractors to get the work done as promised will all factor in to your schedule. Work closely with your contractors to develop a timeline and figure out which activities can be bundled to keep your schedule intact.

**DON'T TRIP**
on your **SHOESTRINGS**

It can't be stressed enough: don't try to cram your building project into a too-tight schedule. Doing so will virtually guarantee you'll end up taking much longer than if you factored in reasonable lead times and padded the schedule between contractors. For example, if your drywall contractor runs longer than expected, your tiling contractor may go off and do another job while he's waiting for yours. That could mean that he'll end up being weeks later than you had originally wanted him. That, in turn, will hold up any activities—such as installation of bathroom fixtures or kitchen cabinets or appliances—that are waiting on tile to be completed. Then the cabinet contractor may go off and take care of another job, pushing you back another couple weeks.

Avoid the domino effect. Set a reasonable schedule, and remain as flexible as you can for delays and changes.

## Love Your Neighbors

If you're building in a residential area, before you start your home building process, check out the etiquette recommendations the National Association of Home Builders Remodelers Council has to get off on the right foot with your neighbors:

- **Keep your neighbors informed.** Let your neighbors know well in advance about your building plans, and keep them apprised of progress, detail by detail. Tell them when work will begin, the approximate completion date, what work will be done, and whether workers might have to come onto their property. If delays arise, promptly contact your neighbors to inform them of the revised schedule.

- **Power down.** Be sure noisy power tools are only used during standard business hours. Reasonable hours are between 8 A.M. and 5 P.M.

- **Keep on truckin'.** Inform your neighbors of any large trucks entering the neighborhood, and ask subcontractors to park on one side of the street only. Try your best to have materials dropped off in your driveway or yard rather than in the street, and keep your yard as tidy as possible. Watch for debris that might find its way onto your neighbors' yards, especially with roofing projects.

- **Dump it.** Remove dumpsters as quickly as possible. If you have room left in your last dumpster, invite neighbors to dump anything they might have lurking in their garage that needs tossing. It doesn't cost you anything more and can be a great convenience for them.

- **Talk it out.** If your neighbors are unhappy with an aspect of your project, promptly visit them to apologize. Consider bringing a peace offering. When your project is complete, show your neighbors your appreciation by throwing them a party or some other form of appreciation. You can thank them for their patience and proudly show off your new house.

# Managing the Project

Whether you're working on your own or with a contractor/manager, you are going to have to be involved in the management of your project. By staying involved, you'll be sure your work gets done in a timely manner and you'll be more likely to avoid costly mistakes.

### Checking Your Schedule

Although you'll be checking your schedule daily to be sure deliveries happen when expected and contractors are working on the agreed-upon dates, you should also take time out frequently to look at the big picture of your schedule. About every 2 weeks, or at least once a month, review your schedule to see if it needs to be modified. Can you move up anything? Will you need to push back any activities?

Be sure you're on-site the day a job is finished. You should carefully inspect the completed work, preferably when the contractor is there so he or she can answer any questions you might have.

Be careful about getting too friendly with your contractors. Sure, they're great people, but save the socializing for after the job is done. You want to keep your relationship on a strictly professional level—you're the boss, and your contractor is an employee. That's not to say you shouldn't be pleasant or fair. However, we have seen firsthand how difficult it can be to point out problems with projects after a contractor has become a friend.

Finally, it's important to pay promptly, but don't pay your contractors their final fees until after the work is done to your satisfaction and you've received lien waivers from all contractors and subcontractors. Your money is leverage to get corrections done to the work, if need be. After you've paid your final bill, that leverage is gone.

### Setting Your Budget

Every home will have its own budget list with different accompanying costs. Much of this will be determined by the decisions you make throughout the process.

Your budget will be a living document that will change many times over the course of your project. So although it's important to have parameters for each line item, it's also important to remain somewhat flexible about them.

## GO FIGURE

Now it's time to take your estimate sheet and turn it into a firm budget. Start with the amount of financing you have available to you. (Get this figure from your mortgage broker, adjusting it if you think it's more debt than you want to take on.)

Go through the process of home building, and create a category for each step. This format will help you set your budget and compare estimates side by side so you can determine which contractors or options will best fit within your financial parameters. (Hint: If you have access to a spreadsheet or word processing program, you can create a file that is easily updated when you get new estimates or quotes.) **Total financing available (the amount of your loan approval or the amount of financing you can comfortably take on)** _____

|  | Budget | Estimate One | Estimate Two | Estimate Three |
|---|---|---|---|---|
| Land | $_____ | $_____ | $_____ | $_____ |
| Surveying | $_____ | $_____ | $_____ | $_____ |
| House plans and architectural services | $_____ | $_____ | $_____ | $_____ |
| Clearing and excavating | $_____ | $_____ | $_____ | $_____ |
| Foundation masonry | $_____ | $_____ | $_____ | $_____ |
| Framing: | | | | |
| Labor | $_____ | $_____ | $_____ | $_____ |
| Materials | $_____ | $_____ | $_____ | $_____ |
| Roofing and gutters: | | | | |
| Labor | $_____ | $_____ | $_____ | $_____ |
| Materials | $_____ | $_____ | $_____ | $_____ |

| | Budget | Estimate One | Estimate Two | Estimate Three | |
|---|---|---|---|---|---|
| Windows and exterior doors | $_____ | $_____ | $_____ | $_____ | |
| Siding: | | | | | |
|   Labor | $_____ | $_____ | $_____ | $_____ | |
|   Materials | $_____ | $_____ | $_____ | $_____ | |
| Heating, ventilation, and cooling | $_____ | $_____ | $_____ | $_____ | |
| Additional ductwork (central vacuum) | $_____ | $_____ | $_____ | $_____ | |
| Plumbing | $_____ | $_____ | $_____ | $_____ | |
| Electric | $_____ | $_____ | $_____ | $_____ | |
| Additional wiring (cable, phone, intercom, etc.) | $_____ | $_____ | $_____ | $_____ | |
| Insulation | $_____ | $_____ | $_____ | $_____ | |
| Drywall | $_____ | $_____ | $_____ | $_____ | |
| Built-ins | $_____ | $_____ | $_____ | $_____ | |
| Fireplace | $_____ | $_____ | $_____ | $_____ | |
| Finishing (trim, doors, paint, and wallpaper) | $_____ | $_____ | $_____ | $_____ | |
| Septic system or sewer hook-up | $_____ | $_____ | $_____ | $_____ | |
| Well or public water hook-up | $_____ | $_____ | $_____ | $_____ | |
| Flooring | $_____ | $_____ | $_____ | $_____ | |
| Kitchen cabinets and fixtures | $_____ | $_____ | $_____ | $_____ | |
| Kitchen countertops | $_____ | $_____ | $_____ | $_____ | |
| Kitchen appliances | $_____ | $_____ | $_____ | $_____ | |
| Bathroom cabinets and fixtures | $_____ | $_____ | $_____ | $_____ | |
| Bathroom countertops | $_____ | $_____ | $_____ | $_____ | |
| Utility room cabinets and fixtures | $_____ | $_____ | $_____ | $_____ | |
| Utility room appliances | $_____ | $_____ | $_____ | $_____ | |

|  | Budget | Estimate One | Estimate Two | Estimate Three |
|---|---|---|---|---|
| Paving | $_____ | $_____ | $_____ | $_____ |
| Patio or deck | $_____ | $_____ | $_____ | $_____ |
| Driveway | $_____ | $_____ | $_____ | $_____ |
| Walkways | $_____ | $_____ | $_____ | $_____ |
| Landscaping | $_____ | $_____ | $_____ | $_____ |
| Permits and inspections | $_____ | $_____ | $_____ | $_____ |
| Other: | | | | |

_____

_____

_____

**Total budget:** _____

Refer to and update this budget sheet often. It's critical that you monitor every aspect of the project cost to ensure you don't run out of money.

## Dollar-Saving Do's and Don'ts

- Use your supply house or contractors to compile free or low-cost take-off lists.

- Always keep resale value in mind when you make decisions about where to cut corners.

- Watch out for hidden inspection and permit costs.

- Don't take your eye off the bottom line. Keep your budget tight, and monitor it regularly to make sure you're within your financial boundaries.

# 14

# THE HOME STRETCH

Now that you know how to take your dream house from a shadowy idea in your head to a beautiful reality, you're almost finished. But first, we need to let you in on a few secrets, tips, and other things people don't usually tell you.

Depending on where you're living during the construction process, you'll need to deal with the transition, which may be timing the end of your rental or selling your current home to coincide with your move-in date.

## Releasing Your Rental

If you're renting your current home, look at your lease to determine what your responsibility is when you want out. Some leases are month-to-month, so you can simply give 30-days notice of your intent to move and you're not responsible for paying rent after that. Some longer-term leases have a 60- or 90-day notification clause, where if you notify the landlord of your intent to leave, giving that much notice, the lease will be terminated at the end of that period. If you don't have such an "out" clause, you need to negotiate with your landlord and see if you can come to an equitable agreement.

Most landlords will work with you if you give them plenty of time to re-rent the property. The tricky part here is that home construction projects can frequently run much longer than expected, so be sure either your landlord will work with you on timing or you have a backup plan in case your move-out date and your move-in date don't coincide. Some people stay with relatives, take a short-term rental, or take advantage of long-term-stay hotels (which usually means staying for longer than a week or two).

## Selling Your Home

If you currently own a home, you have two options: keep it and rent it, or sell it. If you want to keep it and rent it, you should be able to find plenty of books about being a landlord. You'll need liability insurance, and you'll need to follow the renting laws and regulations in your state.

If you're going to be selling your home, you'll want to do what you can to maximize its appeal and its value so you get the best possible price as quickly as you can. Throughout this process, you have a number of considerations.

### Should You Go It Alone?

In Chapter 3, we discussed the advantages and disadvantages of using a Realtor to find a piece of property. Even if you decided to go it alone to shop for land, when selling your home you may want to revisit your decision to do it yourself in light of the complexity involved.

Of course, the big consideration in using a Realtor is that 6 percent of your selling price will get lopped off the top and go to the brokerage. However, a real estate agent performs a number of important services that can make the expenditure worthwhile, including the following:

- Analyzing comparable properties and helping you set a price
- Listing your property in databases accessible to other real estate agents
- Marketing and advertising your property to prospective buyers
- Showing your home to prospective buyers
- Scheduling appointments with inspectors and appraisers and, sometimes, being on hand to open the home to them
- Helping you negotiate purchase offers
- Acting as a resource to answer questions and give you information throughout the process

Finding a reputable real estate agent can make the process go more smoothly. Ask for references and ask the agent questions such as the following:

- How many years have you been in business?
- How long have you been selling locally?
- How many houses did you sell in the past year?
- What is your commission? (Remember, this is often negotiable!)
- If I were to work with you, how would you market my house?
- Will you be on hand to show the house and make arrangements for closing inspections?
- Can you give me at least four recent references?

If you do choose to work with a real estate agent, be sure to check references. Also check with your local Better Business Bureau to ensure that complaints haven't been filed against the agent you're considering working with.

### Sprucing Up Your Home

When you're ready to put your home on the market, you should focus on some quick tips to make your home look the best it can. Don't spend big bucks on major renovations—they're not likely to return the investment. Instead, do the following to make your home look great:

- **Clean from top to bottom.** Do a thorough cleaning job, and keep the home neat during the sales period. You may get last-minute calls for the house to be shown.

- **Unload closets.** Nothing makes closets look smaller than when they're crammed full of clothing and junk. Clear out your closets, and straighten them up to make them look roomier and alleviate storage-space concerns from prospective buyers.

- **Paint.** If your walls are in bad shape, throw on a coat of neutral-colored paint. Light colors usually make rooms look bigger and brighter.

- **Go outside.** Take an afternoon to clean up your home's landscaping, making it look as manicured as possible. If your home's exterior is in bad shape, power wash it. If you must, take some paint to the exterior; but again, home-owners may want to paint their own colors, so it's usually not a good investment to spend much time and expense doing a major exterior remodel.

- **Clean carpets.** Have your carpets steam cleaned to make them look the best they can.

- **Make it bright.** Open the curtains. If the weather is temperate and you aren't near any particularly noisy roadways or other unpleasant sounds, open your windows. For evening visits from prospective buyers, be sure the house is brightly lit.

- **Smell well.** Use wall outlet air fresheners or potpourri to make your home smell great. If you know someone's coming, light a few fragrant candles (cinnamon or other baking smells work great in the kitchen). You could also brew a fresh pot of coffee or bake cookies, letting these homey scents fill your house right before buyers arrive. Smell is linked more strongly to memory than any other sense, so make your home smell great, and you'll have a better chance of standing out.

Overall, keep your home neat and clean. Keep pets in an area where they won't bother prospective buyers. Above all, be sure you remain as flexible as possible in showing your home.

### Timing the Sale

How long will it take to sell your home? That depends on the market, your price, and other mysterious factors that seem to change from sale to sale. We've seen homes that we thought would be a quick sale take months and months, whereas homes we thought were overpriced went in a matter of days.

You can get a better, although not perfect, idea of how quickly homes are selling by speaking with local real estate agents and watching homes for sale in your neighborhood or surrounding areas.

What if you put your home on the market and it sells in just a matter of days—and you still have months left to go on the construction project? Don't worry. It usually takes 2 to 3 months to close a loan. Plus, you may be able to negotiate with the buyers to delay the closing until you're closer to your finish date. If you don't want to wait—and take the chance that something happens to the deal you negotiated—you can instead close the deal and make an arrangement to rent the property from the new owners for a specific amount of time. You can either have the rent deducted from the selling price or pay it monthly. We recommend the latter, because it might be difficult to be reimbursed if you move out early.

If your house is still on the market when you're ready to move, you can apply for a bridge loan, which is intended for just this type of situation. This loan is intended to be a short-term loan that finances both properties concurrently while you try to sell your previous residence. It is refinanced when you do sell that house.

### Setting the Price

When you're working on setting a price for your home, look at comparable sales available from local real estate agents who may give you them for free in exchange for a chance at listing your home. Price your home about 10 to 15 percent higher than what you need or want to get for it. That leaves you room to negotiate the price.

If you price your home too high, you'll likely have a problem selling it. Price it too low, and you could potentially lose thousands of dollars. Think about the following when pricing your home:

- Location
- Market conditions
- Number of similar homes on the market in your area
- Quality of your school district
- Age of your home
- Amount of property you own

- Value-adding features such as a pool, spa, etc.
- Extras included in sales price such as light fixtures, appliances, etc.

A licensed real estate appraiser—someone who assesses home values for banks and mortgage companies—can also give you a good sense of the value of your home. An appraisal can cost you a few hundred dollars, but it will give you the best sense of your home's best price. This may be a worthwhile investment if you want to confirm what your home is worth in the marketplace through a lender's eyes.

### Finding the Right Buyer

If you're working with an agent, he or she will handle qualifying buyers. However, if you're selling your home solo, you can weed out the tire-kickers by only dealing with buyers who are prequalified and have a letter from a lender stating as such. (Again, remember the difference between prequalified and preapproved, the latter of which is not binding; see Chapter 2 for more information.)

You should discuss with the buyer the timeline that he or she will need to close and determine whether that coincides with your own needs. If your buyer wishes to close quickly, you may need to find alterative housing. However, you can also negotiate to delay the closing until your new home is completed. You can also close on the home and then rent it back from the new owners so you don't have to incur additional moving expenses.

## Getting Ready to Move

Selling means moving—and moving can be a big and expensive job. You can either take on the move yourself or hire a moving company to do it. If you're moving yourself, cash in favors from family and friends to help you. Most areas have truck rental companies from which you can rent a vehicle large enough to move your stuff in one or two trips. Be sure to reserve the vehicle in plenty of time. Plus, check online and in your local newspaper for coupons that may save you money on the vehicle.

### Finding the Right Moving Company

If you hire a moving company, it's important to hire a reputable one. A few years ago, a number of news reports detailed disreputable companies selling their services cheaply online and then holding clients' possessions "hostage" until the client paid more money. No one needs those headaches. To be sure you choose a reputable company ...

- Get an estimate in writing for the job, including how long it will take and all fees and charges, including expected gratuities.
- If you can, meet the company representative in person and have him or her examine the property to be moved to ensure the estimate is accurate.
- Check the Better Business Bureau for any complaints against the company, and get—and check—references.
- Inform the company of any circumstances that may make the job difficult, such as moving exceptionally large furniture, the length of staircases, whether movers will need to use an elevator, etc.

### Planning for Your Move

Moving a household requires quite a bit of planning, and it helps to have a timeline. Here are the main tasks you need to address when you move:

**Three months before:**

- Begin purging anything you no longer need—donate old clothes and toys to charity, clear out things you don't need by giving them away or throwing them out. The less you have to move, the less moving will cost.

- Interview, select, and book your moving company.

- Make a list of everyone who needs to know about your move, including family, friends, your employer, your child's school, the IRS, your newspaper carrier, magazines to which you subscribe, and whoever else needs to know where you are.

- Check your homeowner's insurance policy to see what, if any, provision is made for items damaged or lost during the moving process.

**Two months before:**

- Start filling out change-of-address cards.

- Check to see if you need any sort of moving permit.

- If you're moving a long distance, decide what arrangements will be made for pets.

- Properly dispose of items that can't be moved, such as flammable materials, paint, motor oil, or antifreeze.

- Want to make some cash on your move? Have a garage sale! Or start selling stuff on eBay.

- Check with your tax adviser to see if any of your moving expenses are tax deductible.

**One month before:**

- If you're moving yourself, start collecting boxes from friends and neighbors, as well as moving tape, rope, markers, newspapers, and other materials you'll need.

- If you're using a moving company, plan which things will be moved by them and which will be moved by you.
- Head to the local post office, and drop off that change-of-address form so your mail will be forwarded.
- Send change-of-address cards to family and friends. Notify others who need to know you're moving.
- Get your car serviced and ready for the move.
- Call to confirm movers—whether they're volunteers or hired help.

**Two to three weeks before:**

- Notify utility companies to transfer phone, gas, electric, cable, and the like.
- Arrange for baby- or pet-sitting for moving day, if you need it.
- Start packing, if you'll be doing that yourself.

**Two to three days before:**

- Pack valuables and anything the movers won't be taking.
- Drain gas from and clean vehicles that will be moved on the truck, such as lawn mowers, motorcycles, and the like.
- Be sure you have ample boxes, tape, packing materials, markers, and whatever else you'll need to pack.
- Return video rentals and library books.

**The day before:**

- Pick up rental truck or confirm movers.
- Fill up your car.
- Get some sleep.

**Moving day:**

- Label boxes clearly so they can be delivered to the proper rooms in the new house.
- Have packing materials readily available.
- Keep children and pets out of the movers' way.
- If you have family and friends helping you move, make arrangements for drinking water and food for them.
- Enjoy your new home!

## Saving Money on Moving

It's a big job, but you can save money on moving:

- **Do it yourself.** Obviously, if you don't hire movers, you'll save anywhere from $1,500 to $5,000 or more, depending on your move, how much stuff you have, how far away you're moving, and the like.
- **Do part of it yourself.** By packing your own boxes, you could save as much as one third or more.
- **Purge.** As we said earlier, get rid of things you don't use or don't need to save money on the move—the less stuff you have, the less money you'll spend.
- **Materials.** If you're moving yourself, collect used boxes from movers or friends. Use newspapers as a cheap wrapping material to protect items in boxes, and use old blankets to wrap furniture and protect it from scratches.
- **Go online.** Check out the U.S. Postal Service website, Welcomewagon.com, and Valpak.com for moving-related coupons that could save you money on movers, truck rentals, and items you'll need for your new home.

# Keeping the Momentum Going

Face it: something is going to go wrong during the project. Building a home is filled with a million tiny details, all of which have the potential to become a problem. Late deliveries, contractors who don't show up, a header that's framed 2 inches too low—these are all real-life experiences and good reasons why you need to stay on top of the project from first dig to finishing touches.

Sometimes, however, this all can get stressful and discouraging. But running out of steam midway through your project isn't a good thing.

### Dealing With Stress

Throughout this stressful process, be sure to take care of yourself. Try to get enough sleep and tackle problems one by one instead of looking at the whole monumental effort involved in building a home and moving to a new residence. Try some proven stress-management techniques such as deep breathing and writing down your concerns in a journal.

Most of all, don't let the stress of all this take its toll on your family. When you feel overwhelmed, take a walk or talk to a friend. Remember, you're doing this to have a beautiful new home for your family, and your compensation for the stress and headaches is the big chunk of change you're saving by using the strategies in this book.

### Unexpected Expenses

When unexpected expenses arise, you need to find a way to balance your budget. Don't fool yourself into thinking that a few hundred dollars here and a thousand dollars there isn't a big deal—they can be if you don't watch the numbers. When something goes over budget in one area, you need to immediately cut back your budget in another area.

If you can't find cuts you can make in your budget, use the principles we've outlined in this book about finding less-expensive materials or delaying some projects until after you move into the house. Even though you might not like it, it's a much better decision to put in laminate countertop and hold off on installing most of your molding, than it is to go full force and find out that you have no money left for flooring or that you can't pay your contractors because you've run out of dough.

### If Money Gets Tight

If the worst does happen and you find that you're running out of money, speak to your lender as soon as possible. Lenders don't like it when people come back to them for more money, but they don't want to see projects such as this fail, either. When that happens, it's a big expense for them to foreclose, so they'll likely work with you, depending on how far over budget you are, the amount of credit for which you qualify, and how far along the job is.

## A Final Word on Saving Costs

We've covered many cost-savings tips in these pages. Some just don't fit into a narrow category or are so important they're worth repeating:

- **Network.** Get to know people in your neighborhood, in your local business associations, at church, at your child's school, etc. The more people you know, the more likely you'll meet someone in the construction trades. People you know are more likely to give you better prices and good advice.

- **Call in favors.** If you have friends in the business, don't be afraid to cash in favors. Ask for advice and help from friends. After all, this is likely to be one of if not *the* biggest project you've ever taken on. That's what friends are for.

- **Don't be afraid to negotiate.** We've said it before, but it's important: You can't be afraid to ask for a better price or better terms.

- **Go online.** Your computer can be a great place for savings. You can comparison shop for prices and often find better prices online. Just be sure the shipping cost doesn't eat up your savings. Another smart strategy is to find the best price online and then ask your local retailer to match it.

  We also purchased two 10 percent off coupons—one for Lowe's and one for The Home Depot, on eBay for $5. We used them to save more than $1,200 by grouping our purchases and taking advantage of the discounts.

  In addition, websites like www.coolsavings.com, www.shopzilla.com, www.jumpondeals.com, and www.netdeals.com have coupons free for the printing. A good way to save on what you need is to put the item which you need to buy and "coupon" or "promotional code" into your favorite search engine. For instance, try "kitchen cabinets, coupons." You're likely to turn up a bunch of coupons and special offers on the items for which you're shopping.

- **Go for bulk discounts.** Ask your new neighbors what home improvements they might need, such as new roofs, driveway resurfacing, fencing, or new swimming pools, and then negotiate a better price with the contractor by using the volume as leverage. You'll get a better price if you can deliver three fencing jobs in the same area that can be completed consecutively than if you just have your own.

- **Watch out for budget-busters.** The often-overlooked details, such as cabinet hardware, door locks, and the like, can quickly add up. So can inspections and permit fees that weren't included in the budget or necessary finishing materials that weren't estimated. Be sure you meticulously document every item you'll need to build and finish your home.

- **Delay some projects.** If you run into budgeting problems, delay some of the least-pressing jobs to give your budget some breathing room.

## Moving On

Thank you for letting us share these ideas with you. We hope our experiences have been helpful and will save you time, hassle, headaches, and, most important, money!

Good luck with your building project. We wish you many happy years in your home.

## Dollar-Saving Do's and Don'ts

- Give yourself plenty of time to sell your existing home or advise your landlord of your intent to move. More time will give you more flexibility.

- Use simple methods to spruce up your home and make it more inviting—and possibly sell more quickly.

- Like building your home, moving requires keeping track of lots of details. Make a checklist to ensure nothing falls through the cracks.

- Be prepared for unexpected expenses. If money gets tight, you need to find ways to balance your budget or speak with your lender.

- When it comes to saving money, be creative and pull out all the stops. Negotiate, cash in favors, use the Internet— the benefits go directly into your pocket and pad your bottom line.

# Appendix A
# GLOSSARY

**adjustable rate mortgage (ARM)**   A home loan in which the interest rate can change.

**anchor bolt**   A bolt used to secure a wood sill to a house's foundation.

**asphalt**   A solid or semisolid mixture used in paving, roofing, and waterproofing.

**awning window**   A window that connects to the house with a hinge at the top and opens away from the house by turning a crank.

**backfill**   Soil used to refill an excavated area.

**backsplash**   A small vertical border that protects the wall behind a sink, stove, or countertop from moisture.

**baluster**   A supporting post of a handrail.

**balloon framing**   A method of building the structure of a house where extra-long vertical posts extend from the foundation to the roof.

**bay window**   Any window space projecting outward from the walls of a building, either square or polygonal in plan.

**bearing wall**   A wall that supports the weight of the structure.

**blueprint**   The architectural drawings and specifications of a home design.

**blown-in insulation** Insulation that is installed by using air power to blow small pieces of insulation into an area.

**bow window** An angled combination of windows in 3-, 4-, or 5-lite configurations. The windows are attached at 10-degree angles to project a more circular, arced appearance.

**bridge loan** A loan that finances two properties for a short term until one of the properties is sold.

**building code** Municipality-required regulations and guidelines for building a structure.

**casement window** A window that is attached to the home on one side by a hinge and extends outward from the home by turning a crank.

**circuit** A closed path of electrical flow.

**circuit breaker** A device that causes a closed circuit to be broken, ending the flow of electricity, usually when the circuit is overloaded.

**clapboard** A form of wood house siding made of overlapping horizontal boards.

**compressor** Part of a heating or cooling system that increases the pressure of air flow.

**concrete** A hard material made by mixing sand, gravel, cement, and water.

**condenser** A heat exchanger in which the refrigerant, compressed to a hot gas, is condensed to liquid by rejecting heat.

**construction loan** A loan that finances a homebuilding project for the term of the project.

**contract** A written agreement between two parties to conduct business.

**cornice** A decorative framework used at the top of a window casing.

**chlorinated polyvinyl chloride (CPVC)**   A type of plastic pipe often used for water supply.

**credit report**   A history of an individual's credit access and usage over a period of time.

**credit score**   A numerical ranking, based on an individual's past credit history, that can affect access to and terms of future credit. Also called a FICO score.

**crown molding**   Decorative interior trim, usually used at the joint where the ceiling meets the wall.

**debt-to-income ratio**   The amount of loans, credit cards, and other monies owed in relation to the amount of money an individual makes.

**deck**   An exterior structure that extends usually from the back of a home, often providing additional living space.

**disbursement**   An installment payment from the monies available from the construction loan to finance the construction project. Also called a draw.

**double-hung window**   A window with two sliding segments that move up and down.

**downspout**   A vertical pipe affixed to the exterior of a home and to a gutter that leads water off the roof and away from the home.

**drain-waste-vent (DWV) system**   The portion of the plumbing system that takes waste water and material from the home and delivers it to the sewer system or septic tank.

**draw**   *See* disbursement.

**drywall**   An interior wall covering, usually gypsum board, that is affixed to framing studs. Also called wall board or sheetrock.

**ducts**   Piping or tubing that delivers air throughout a home.

**elevation**   A drawn view of the exterior of a home from a particular angle.

**equity**   The amount of money a home is worth above and beyond the debt owed on the home.

**excavation**   The removal of debris, soil, and barriers to a home's construction and the digging of the home's foundation.

**facade**   The exterior front of a house.

**fascia**   A board that runs horizontally across the ends of the roof rafters ends, creating the "edge" of the roof.

**FICO score**   *See* credit score.

**fixed window**   A pane of glass that does not open and remains stationary.

**fixed-rate mortgage**   A home loan in which the interest rate stays fixed throughout the life of the loan.

**floating floor**   A flooring material that's laid over the subfloor and that is not secured with an adhesive.

**flue**   The space within a chimney that allows smoke, gas, or fumes to be vented upward and out of the home.

**foam insulation**   Insulation that is either a rigid board affixed to the outside of the home or a liquidlike substance that can be sprayed into crevices and then expands.

**footings**   Concrete structures that support a home's foundation.

**foundation**   The supporting structure of a home.

**framing**   The interior part of a home's structure, usually constructed of wood and/or steel, which creates the home's walls, floors, door and window openings, and roof.

**gable**   The vertical triangular wall between the sloping ends of a gable roof.

**geothermal heating system**   A heating system that forces air into the ground, where it is heated using the earth's temperature.

**girder**   A large wood or steel beam that is the main support of a structure's weight.

**green board**   A form of water-resistant drywall usually used in bathrooms and other very humid areas.

**ground fault interrupter (GFI)**   A device which interrupts the electric circuit when the current to ground exceeds a predetermined value.

**gutters**   U-shaped channels affixed horizontally along sections of the roof, that catch and deliver rainwater into a downspout, which then takes the water to the ground and/or away from the house.

**header**   A piece of lumber installed horizontally over an opening in the framing, such as a window or door, to support the weight above the opening.

**heat pump**   A device that moves hot air from one place to another.

**HVAC (heating, ventilating, and air-conditioning) system**
The home system that regulates interior temperature and air flow.

**insulation**   A material, usually fiberglass, cellulose, or foam, that is installed along the exterior walls and flooring of your structure to maintain temperature more effectively within the structure.

**interest**   The amount of money a lender charges a borrower, above and beyond the amount borrowed.

**joint**   The point at which two pieces of building material come together. Joints can be found where the ceiling meets the floor or where two pieces of pipe are affixed.

**joist**   A horizontal beam used to support floor and ceiling loads and supported in turn by larger beams, girders, or bearing walls.

**level**   Perfectly horizontal. Also refers to the tool used to determine if something is perfectly horizontal or plumb.

**lien**   A financial claim against a property, usually filed when monies are owed.

**lien waiver**   A document that certifies that a contractor has been paid for services and materials delivered and cancels the right to file a lien against a property.

**loan-to-value ratio**   The proportion of money lent on a property to the actual market value of the property.

**main drain**   The primary waste pipe that moves waste material from the home to the sewer or septic system.

**main water supply**   The primary pipe through which water is delivered to the home.

**masonry**   Concrete, brick, and stone work done on the home, including the foundation, footings, and any decorative or structural brick or stone work.

**molding**   Decorative finished wood or composite pieces that cover joints between walls and ceilings or walls and floors. Also used around windows to cover gaps between drywall and window jambs. Also called trim.

**mortgage**   Money loaned to an individual or couple to purchase a home.

**nail plate**   A small piece of metal that protects wires and pipes within walls from being punctured by nails or screws.

**outlet**   A receptacle that delivers electricity to appliances and devices when they are connected to it through a cord.

**particleboard**   A sheet of building material made of wood particles and glue that are bonded under pressure.

**paving stone**   A concrete or stone piece that interlocks with other uniform pieces to form a flat surface. Often used for patios, driveways, or walkways. Also called a paver.

**pier foundation**   A foundation which uses wood posts instead of a concrete slab to support the structure.

**pile foundation**   A foundation that uses concrete pillars instead of a concrete slab to support the structure.

**plaster**   A mixture of lime or gypsum with sand and water that hardens into a smooth solid and is used to cover walls and ceilings.

**plate**   A horizontal piece of wood or metal that anchors studs to a floor or ceiling.

**platform framing**   A method of framing in which each floor supports the one above it. The first floor is supported by the foundation and bearing walls. The second floor is supported by the first floor and bearing walls, etc.

**plumb**   Perfectly vertical.

**plywood**   A piece of wood made of three or more layers of veneer joined with glue and usually laid with the grain of adjoining plies at right angles.

**point**   One percent of the loan value, often charged as a fee by the lender.

**polyvinyl chloride (PVC)**   A plastic pipe used in residential plumbing systems.

**pounds per square inch (psi)**   A measurement of the pressure a material exerts on the walls of a confining vessel or enclosure.

**preapproval**   An estimate of the amount of money for which a mortgage applicant will be approved, based on preliminary information.

**prequalification**   A commitment from a lender to lend a certain amount of money to a mortgage applicant, based on an application and verification of information.

**pressboard**   *See* particleboard.

**principal**   The amount of money owed to the lender, not including interest, PMI, or other fees.

**private mortgage insurance (PMI)** Insurance required by lenders in situations where the borrower has less than 20 percent equity.

**ridge vent** An opening in the highest point of the roof that allows air flow.

**rise** The vertical height from one step to the next.

**risers** The vertical boards on a stairway that support the treads.

**rough-in** The initial placement and installation of the plumbing system before it is connected and sealed to carry water and other liquids.

**run** *See* tread.

**R-value** A unit of thermal resistance used for comparing insulating values of different materials. The higher a material's R-value, the greater its insulating properties and the slower heat flows through it.

**sheathing** A protective covering consisting, for example, of a layer of boards applied to the studs and joists of a building to strengthen it and serve as a foundation for a weatherproof exterior.

**sheetrock** *See* drywall.

**sill** Boards bolted to the foundation to which the framing is attached.

**slab** A thick, solid block of concrete.

**soffit** The underside of a roof's overhang.

**solder** An alloy that melts at a fairly low temperature and is used for making permanent connections between pipes and wires.

**splashblock** A tray located at the bottom of a downspout that directs water away from a home's foundation.

**stucco** A plaster compound applied while soft to cover exterior walls or surfaces.

**subfloor**   Boards or plywood that lie on joists over which finish flooring material is placed.

**switch**   A device that controls the flow of electricity to a light or socket.

**take-off list**   A list of materials necessary to build a house.

**tongue and groove**   Lumber that fits together through inter-locking grooves and extensions along the edges.

**tracking pad**   A rough, gravel-covered route that construction vehicles use to access a construction site.

**tread**   The horizontal part of a stair on which people step. Also called the run.

**trim**   *See* molding.

**truss**   A premanufactured structure used for constructing roofs.

**U-value**   A measurement of heat retention used for glass.

**vapor barrier**   A plastic sheet that prevents moisture exposure. Usually used to protect walls, floors, and/or ceilings.

**vinyl siding**   Strips of vinyl that are used to cover the exterior of a home.

**wall board**   *See* drywall.

**workers' compensation insurance**   Insurance carried by employers that covers any liability incurred when an employee is injured on the job.

# Appendix B
# RESOURCES

## Associations

**Air Conditioning Contractors of America**
2800 Shirlington Road, Suite 300
Arlington, VA 22206
703-575-4477
Fax: 703-575-4449
www.acca.org

**American Association of Subcontractors**
1004 Duke Street
Alexandria, VA 22314
703-684-3450
Fax: 703-836-3482
www.asaonline.com/

**American Concrete Institute (ACI) International**
PO Box 9094
Farmington Hills, MI 48333
248-848-3700
Fax: 248-848-3701
www.aci-int.org

**American Concrete Pavement Association**
1010 Massachusetts Avenue, NW, Suite 200
Washington, DC 20001
202-842-1010
Fax: 202-842-2022
www.pavement.com

**American Hardware Manufacturers Association**
801 North Plaza Drive
Schaumburg, IL 60173-4977
847-605-1025
Fax: 847-605-1030
www.ahma.org

**American Institute of Architects**
1735 New York Avenue, NW
Washington, DC 20006-5292
202-626-7300 or 1-800-AIA-3837
(1-800-242-3837)
Fax: 202-626-7547
www.aia.org

**American Institute of Building Design**
2505 Main Street, Suite 209B
Stratford, CT 06615
1-800-366-2423
Fax: 203-378-3568
www.aibd.org

**American Institute of Steel Construction, Inc.**
One East Wacker Drive, Suite 3100
Chicago, IL 60601-2001
312-670-2400
Fax: 312-670-5403
www.aisc.org

**American Institute of Timber Construction**
7012 S. Revere Parkway, Suite 140
Englewood, CO 80112
303-792-9559
Fax: 303-792-0669
www.aitc-glulam.org

**American Iron and Steel Institute**
1140 Connecticut Avenue, Suite 705
Washington, DC 20036
202-452-7100
www.steel.org

**American Lighting Association**
PO Box 420288
Dallas, TX 75342-0288
1-800-274-4484
www.americanlightingassoc.com

**American National Standards Institute**
1819 L Street, NW (between 18th and 19th Streets), 6th Floor
Washington, DC 20036
202-293-8020
Fax: 202-293-9287
www.ansi.org

American Plywood Association and
Engineered Wood Association
7011 So. 19th
Tacoma, WA 98466
253-565-6600
Fax: 253-565-7265
www.apawood.org

American Society of Heating,
Refrigerating and Air-Conditioning
Engineers, Inc.
1791 Tullie Circle, NE
Atlanta, GA 30329
404-636-8400
Fax: 404-321-5478
www.ashrae.org

American Solar Energy Society
2400 Central Avenue, Suite A
Boulder, CO 80301
Fax: 303-443-3212
www.ases.org

American Water Works Association
6666 W. Quincy Avenue
Denver, CO 80235
303-794-7711 or
1-800-926-7337
Fax: 303-347-0804
www.awwa.org

Architectural Woodwork Institute
1952 Isaac Newton Square West
Reston, VA 20190
703-733-0600
Fax: 703-733-0584
www.awinet.org

Associated Builders and Contractors
4250 N. Fairfax Drive,
9th Floor
Arlington, VA 22203-1607
703-812-2000
www.abc.org

Associated General Contractors of
America
333 John Carlyle Street, Suite 200
Alexandria, VA 22314
703-548-3118
Fax: 703-548-3119
www.agc.org

Associated Soil and Foundation
Engineers
8811 Colesville Road, Suite G106
Silver Spring, MD 20910
301-565-2733
Fax: 301-589-2017
www.asfe.org

Association of Home Appliance
Manufacturers
1111 19th Street, NW, Suite 402
Washington, DC 20036
202-872-5955
Fax: 202-872-9354
www.aham.org

Association of Millwork Distributors
10047 Robert Trent Parkway
New Port Richey, FL 34655-4649
727-372-3665
Fax: 727-372-2879
www.nsdja.com

**Asphalt Institute**
PO Box 14052
Lexington, KY 40512-4052
859-288-4960
Fax: 859-288-4999
www.asphaltinstitute.org

**Asphalt Roofing Manufacturers Association**
Public Information Department
1156 15th Street, NW, Suite 900
Washington, DC 20005
202-207-0917
Fax: 202-223-9741
www.asphaltroofing.org

**Blow-in-Blanket Contractors Association**
1051 Kennel Drive
Rapid City, SD 57702
1-800-451-8862
www.bibca.org

**Brick Industry Association**
11490 Commerce Park Drive
Reston, VA 20191-1525
703-620-0010
Fax: 703-620-3928
www.bia.org

**Carpet Rug Institute**
310 Holiday Avenue
Dalton, GA 30720
*Mailing address:*
PO Box 2048
Dalton, GA 30722-2048
703-875-0634
Fax: 703-875-0907
www.carpet-rug.org

**Cedar Shake and Shingle Bureau**
PO Box 1178
Sumas, WA 98295-1178
604-820-7700
Fax: 604-820-0266
www.cedarbureau.org

**Cellulose Insulation Manufacturers Association**
136 S. Keowee Street
Dayton, OH 45402
937-222-CIMA (937-222-2462) or
1-888-881-CIMA
(1-888-881-2462)
Fax: 937-222-5794
www.cellulose.org

**Concrete Reinforcing Steel Institute**
933 North Plum Grove Road
Schaumburg, IL 60173-4758
847-517-1200
Fax: 847-517-1206
www.crsi.org

The Construction Specifications
Institute
99 Canal Center Plaza, Suite 300
Alexandria, VA 22314
1-800-689-2900
Fax: 703-684-8436
www.csinet.org

Copper Development Association
260 Madison Avenue
New York, NY 10016
212-251-7200
Fax: 212-251-7234
www.copper.org

Energy and Environmental Building
Association
10740 Lyndale Avenue South, Suite
10W
Bloomington, MN 55420-5615
952-881-1098
Fax: 952-881-3048
www.eeba.org

Environmental Protection Agency
Ariel Rios Building
1200 Pennsylvania Avenue, NW
Washington, DC 20460
202-272-0167
www.epa.gov

Hardwood Manufacturers
Association
400 Penn Center Boulevard, Suite 530
Pittsburgh, PA 15235
1-800-373-WOOD
(1-800-373-9663)
www.hardwoodcouncil.com

Home Builders Institute
1201 15th Street, NW,
Sixth Floor
Washington, DC 20005
1-800-795-7955
Fax: 202-266-8999
www.hbi.org

Institute of Electrical and
Electronics Engineers, Inc.
3 Park Avenue, 17th Floor
New York, New York 10016-5997
212-419-7900
Fax: 212-752-4929
www.ieee.org

Insulation Contractors Association of
America
1321 Duke Street, Suite 303
Alexandria, VA 22314
703-739-0356
Fax: 703-739-0412
www.insulate.org

Kitchen Cabinet Manufacturers
Association
1899 Preston White Drive
Reston, VA 20191-5435
703-264-1690
Fax: 703-620-6530
www.kcma.org

**Light Gauge Steel Engineers Association**
1201 15th Street, N.W., Suite 320
Washington, DC 20005
202-263-4488 or 866-465-4732
Fax: 202-785-3856
www.lgsea.com

**National Asphalt Pavement Association**
5100 Forbes Boulevard
Lanham, MD 20706
1-888-468-6499
www.hotmix.org

**National Association of Home Builders**
1201 15th Street, NW
Washington, DC 20005
202-266-8200 or
1-800-368-5242
www.nahb.com

**National Association of Women in Construction**
327 S. Adams Street
Fort Worth, TX 76104
817-877-5551 or
1-800-552-3506
Fax: 817-877-0324
www.nawic.org

**National Concrete Masonry Association**
13750 Sunrise Valley Drive
Herndon, VA 20171-4662
703-713-1900
Fax: 703-713-1910
www.ncma.org

**National Electrical Contractors Association**
3 Bethesda Metro Center, Suite 1100
Bethesda, MD 20814
301-657-3110
Fax: 301-215-4500
www.necanet.org

**National Electrical Manufacturers Association**
1300 N. 17th Street, Suite 1847
Rosslyn, VA 22209
703-841-3200
Fax: 703-841-5900
www.nema.org

**National Hardwood Lumber Association**
6830 Raleigh-LaGrange Road
Memphis, TN 38184-0518
901-377-1818
www.natlhardwood.org

**National Kitchen and Bath Association**
687 Willow Grove Street
Hackettstown, NJ 07840
1-800-843-6522
Fax: 908-852-1695
www.nkba.org

National Paint and Coatings
Association
1500 Rhode Island Avenue, NW
Washington, DC 20005
202-462-6272
Fax: 202-462-0347
www.paint.org

National Roofing Contractors
Association
10255 W. Higgins Road, Suite 600
Rosemont, IL 60018
847-493-7573
Fax: 847-299-1183
www.nrca.net

National Rural Water Association
2915 South 13th Street
Duncan, OK 73533
580-252-0629
Fax: 580-255-4476
www.nrwa.org

National Stone, Sand and Gravel
Association
1605 King Street
Alexandria, VA 22314
703-525-8788 or
1-800-342-1415
Fax: 703-525-7782
www.nssga.org/

National Tile Contractors
Association
PO Box 13629
Jackson, MS 39236
601-939-2071
Fax: 601-932-6117
www.tile-assn.com

National Wood Flooring Association
111 Chesterfield Industrial Boulevard
Chesterfield, MO 63005
1-800-422-4556
Fax: 636-519-9664
www.woodfloors.org

North American Insulation
Manufacturers Association
44 Canal Center Plaza, Suite 310
Alexandria, VA 22314
703-684-0084
Fax: 703-684-0427
www.naima.org

Plumbing-Heating-Cooling
Contractors Association
180 S. Washington Street
PO Box 6808
Falls Church, VA 22040
703-237-8100 or
1-800-533-7694
Fax: 703-237-7442
www.phccweb.org

**Sheet Metal and Air Conditioning Contractors' National Association**
4201 Lafayette Center Drive
Chantilly, VA 20151-1209
703-803-2980
Fax: 703-803-3732
www.smacna.org

**Steel Door Institute**
The Steel Door Institute is managed at:
30200 Detroit Road
Cleveland, OH 44145-1967
440-899-0010
Fax: 440-892-1404
www.steeldoor.org

**Steel Framing Alliance**
1201 15th Street, NW, Suite 320
Washington, DC 20005
202-785-2022
www.steelframingalliance.com

**Timber Framers Guild of North America**
PO Box 60
Becket, MA 01223
Phone and fax: 1-888-453-0879
www.tfguild.org

**U.S. Department of Housing and Urban Development**
451 7th Street, SW
Washington, DC 20410
202-708-1112
www.hud.gov

**Wallcoverings Association**
401 N. Michigan Avenue
Chicago, IL 60611
312-644-6610
www.wallcoverings.org

**Window and Door Manufacturers Association**
1400 E. Touhy Avenue, Suite 470
Des Plaines, IL 60018
847-299-5200 or
1-800-223-2301
Fax: 847-299-1286
www.nwwda.org

**World Floor Covering Association**
2211 East Howell Avenue
Anaheim, CA 92806
714-978-6440 or
1-800-624-6880
Fax: 714-978-6066
www.wfca.org

## Calculators and Ratings

**www.Bankratc.com**
Free mortgage calculators and amortization charts.

**www.consumerreports.com**
Objective ratings on hundreds of home and homebuilding products and appliances.

## Other Handy Resources

**www.aecdaily.com**
Trade construction information and resources, but good information for owner-builders as well.

**www.askthebuilder.com**
Nationally syndicated newspaper columnist Tim Carter answers questions and offers many articles about several homebuilding topics.

**www.codecheck.com**
Check building code requirements in your area.

**www.B4UBuild.com**
Designed for anyone involved in residential construction, this site has informative articles, sample construction schedules and contracts, and a directory of links to useful information and resources.

**www.contractor-info.com**
Information about general contractor requirements in your state. Published by the National Association of Home Builders.

**www.doityourself.com**
A compilation of how-to information, instructions, and community forums where you can ask questions and interact with other homebuilders and re-modelers.

**www.diynet.com**
At the website of the Do It Yourself network, you can find community forums, articles, and downloadable how-to videos for a wide variety of homebuilding projects.

**www.energy.gov**
The U.S. Department of Energy website has both informative money-saving tips and insulation R-value recommendations for various areas of the country.

**www.hammerzone.com**
Articles about home construction.

**www.homebuilding.about.com**
A collection of information and resources hosted on the About.com network of websites.

**www.permitplace.com**
Online access to more than 1,300 county and nearly 3,000 city permit requirements.

**www.taunton.com/finehomebuilding**
From *Fine Homebuilding* magazine, this site offers a lively forum for homebuilders to ask questions and get answers from experienced professionals.

# Appendix C
# SAMPLE DOCUMENTS

Your homebuilding project will generate many forms and documents. We've included a few samples here so you have a good sense of what you'll encounter, including sample elevations and floor plans, a sample contractor agreement, and a sample lien waiver. Remember, it's always a good idea to have your attorney review any legal document prior to use.

# Sample Elevations and Floor Plans

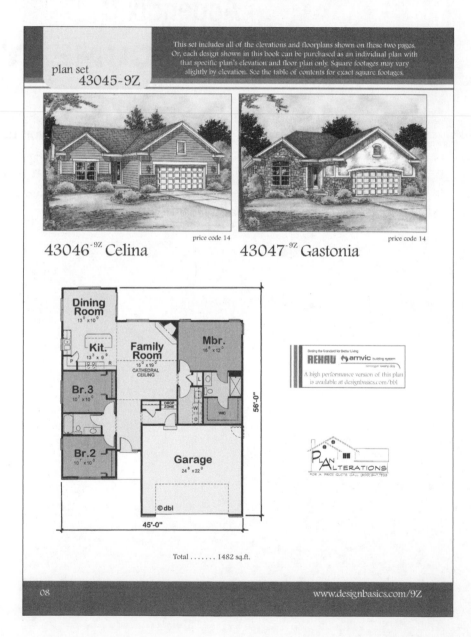

plan set
43045-9Z

This set includes all of the elevations and floorplans shown on these two pages. Or, each design shown in this book can be purchased as an individual plan with that specific plan's elevation and floor plan only. Square footages may vary slightly by elevation. See the table of contents for exact square footages.

price code 14

price code 14

43046-9Z Celina          43047-9Z Gastonia

Dining Room 13⁰ x 10⁰

Kit. 13³ x 9⁹

Family Room 15⁰ x 19⁰ CATHEDRAL CEILING

Mbr. 16⁸ x 12⁰

Br.3 10⁷ x 10⁵

Br.2 10⁷ x 10⁵

Garage 24⁸ x 22⁰

DROP ZONE

WIC

56'-0"

45'-0"

© dbi

REHAU ■ amvic building system

A high performance version of this plan is available at designbasics.com/bb!

PLAN ALTERATIONS

Total . . . . . . . 1482 sq.ft.

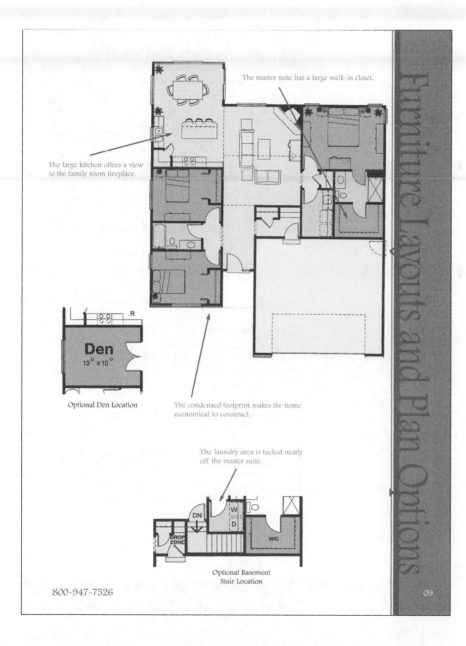

The master suite has a large walk-in closet.

The large kitchen offers a view to the family room fireplace.

**Den**
13⁰ x 10⁰

Optional Den Location

The condensed footprint makes the home economical to construct.

The laundry area is tucked neatly off the master suite.

DN

W
D

DROP
ZONE

WIC

Optional Basement
Stair Location

Furniture Layouts and Plan Options

09

plan set
**43036-9Z**

This set includes all of the elevations and floorplans shown on these two pages. Or, each design shown in this book can be purchased as an individual plan with that specific plan's elevation and floor plan only. Square footages may vary slightly by elevation. See the table of contents for exact square footages.

price code 16

price code 16

**43037**-⁹ᶻ **Limington**

**43038**-⁹ᶻ **Tillamook**

Mbr.
13⁰ x 15⁰

Kit.

Dining Room
9⁴ x 13⁰

WIC

DROP ZONE

W  D

STORAGE

UP

BRM

UP

Family Room
14⁰ x 18⁰
SLOPED CEILING

ENT CENTER

Garage
19⁴ x 28⁰

©dbi

48'-0"

42'-0"

REHAU  amvic building system
Getting the Standard for Better Living
stronger every day
A high performance version of this plan is available at designbasics.com/bbl

Br.3
10⁰ x 12⁰

DN

WIC

Br.2
12⁰ x 11⁰

WIC

Main Level . . 1173
Second Level . . 465
Total . . . . . . . 1638 sq.ft.

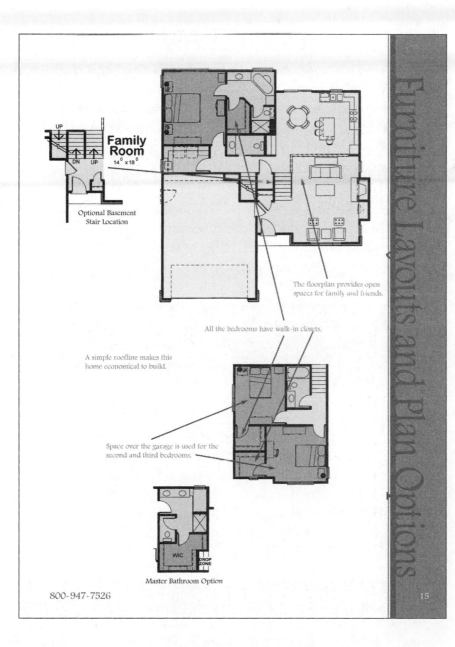

Family
Room
14⁰ x 18⁰

Optional Basement
Stair Location

The floorplan provides open
spaces for family and friends.

All the bedrooms have walk-in closets.

A simple roofline makes this
home economical to build.

Space over the garage is used for the
second and third bedrooms.

WIC   DROP
ZONE

Master Bathroom Option

800-947-7526

15

Furniture Layouts and Plan Options

# Sample Contractor Agreement

Disclaimer: This agreement reprinted with permission of Utahbuild. This sample agreement is intended to help educate homeowners on the types of things they need to agree on with their contractors. Utahbuild does not represent that this sample contract is valid in your area. It is the responsibility of each homeowner to be advised by competent legal counsel regarding these matters.

THIS AGREEMENT is entered into this day of MONTH DAY, YEAR, by and between CONTRACTOR, whose business is located at CONTRACTORS ADDRESS, hereinafter referred to as Contractor, and HOMEOWNER'S NAME, hereinafter referred to Homeowner, whose residence is located at THE HOME'S ADDRESS. Witnesseth, that Homeowner wishes to have home improvement and other services provided at the address given above and Contractor has agreed to provide such services in compliance with the terms and conditions and for the consideration set forth as follows.

### Scope of Work

Contractor shall perform all of the work shown on the Plans and/or described in the Specifications set forth in Exhibit A, attached hereto and incorporated by this reference. The work shall be performed by Contractor on the property located at THE HOME'S ADDRESS. Contractor shall furnish all of the materials, licenses, permits, and any and all other products and services required for the completion of the work under this Agreement.

### Time of Completion

Contractor shall commence the work under this Agreement on or before START DATE, and the work shall be completed on or before COMPLETION DATE. Time is of the essence. The following shall constitute substantial commencement of work under this Agreement.

## Agreement Price

Homeowner shall pay Contractor the sum of PRICE AGREED UPON DOLLARS for all materials and work provided by Contractor under this Agreement, subject to additions and deductions pursuant to authorized change orders made in writing and signed by Contractor and Homeowner.

## Payment Terms

The Agreement price shall be paid as work progresses in accordance with the following schedule of payments.

Homeowner shall pay Contractor the sum of EARNEST AMOUNT DOLLARS upon signature of this Agreement.

Homeowner shall pay Contractor the sum of INTERMEDIATE AMOUNT DOLLARS on Intermediate Date.

Homeowner shall pay Contractor the sum of REMAINING SUM DOLLARS upon completion of all scheduled and agreed work. All work is scheduled to be completed on COMPLETION DATE.

## Change Orders/Modifications to Scope of Work

Any modifications to this Agreement or to the specifications and scope of work detailed in Exhibit A must be agreed to in writing by Contractor and Homeowner. Any and all such modifications must clearly set forth the changes being agreed to and how said changes will impact the Agreement price, any change in Agreement price agreed to in such modification shall be reflected and incorporated into the Agreement price under this Agreement.

## Failure by Contractor to Meet Work Progress Deadlines and Payment by Homeowner

Homeowner and Contractor agree that time is of the essence under this Agreement. Contractor and Homeowner have agreed to the Timeline and Progress Schedule set forth in Exhibit B, attached hereto and incorporated by this reference. Contractor and Homeowner agree that the progress payments set forth in this Agreement shall be made to Contractor upon Contractor meeting the targets set forth and agreed to in Exhibit B. If payment is not made by Homeowner upon Contractor's successful completion of the work set forth in Exhibit B, Contractor may cease work under this Agreement until the amount due has been paid. In the event that Homeowner fails to make a scheduled payment for more than HOMEOWNER'S GRACE PERIOD DAYS after that payment becomes due, such failure shall constitute a material breach of this Agreement. In the event that Contractor fails to meet the deadlines set forth in Exhibit B for a period of CONTRACTOR'S Grace Period Days and a Change Order and/or Modification to this Agreement has not been obtained, such failure shall constitute a material breach of this Agreement.

### General Provisions

In addition to the provisions set forth above, the following general provisions shall also apply:

All work shall be performed and completed in a workmanlike manner and shall comply with all building codes and other applicable laws.

Contractor shall furnish Homeowner detailed plans; drawings; materials lists showing and detailing the shape, size, dimensions, and all equipment and construction materials to be used; a description of all work to be completed; and a description of all materials and equipment to be used or installed and stating the agreed upon AGREEMENT PRICE.

All work shall be performed by duly licensed and legally authorized individuals to the extent required by law.

Contractor shall be fully responsible for payment in full to all subcontractors engaged by in the completion of the work under this Agreement, and Contractor shall remain solely liable and responsible for all work completed under this Agreement.

Contractor shall furnish Homeowner all releases or lien waiver documents for all work performed or materials used in completion of this Agreement at the time that the scheduled payment from Homeowner is due.

Contractor warrants to Homeowner that it is adequately insured for injury to its employees and others incurring loss or injury as a result of the acts of Contractor or Contractors employees or subcontractors.

Contractor shall obtain all necessary permits for the work to be performed under this Agreement at Contractor's sole expense and shall provide proof of such documents to Homeowner prior to commencing the work under this Agreement.

Contractor shall remove all debris and leave the premises in broom clean condition.

Contractor shall not be liable for any delay due to circumstances beyond its control including strikes, natural disaster, or casualty.

## Severability

In case any provision of this Agreement shall be invalid, illegal or unenforceable, such provision shall be construed so as to render it enforceable and effective to the maximum extent possible in order to effectuate the intention of this Agreement; and if such provision shall be wholly invalid, illegal, or unenforceable, the validity, legality, and enforceability of the remaining provisions hereof shall not in any way be affected or impaired thereby.

### Titles and Subtitles

The titles of the Articles and Sections of this Agreement are for convenience of reference only and are not to be considered in construing this Agreement.

### Arbitration

Contractor and Homeowner mutually agree that all disputes arising under this Agreement shall be resolved by binding arbitration in accordance with the rules of the American Arbitration Association.

### Entire Agreement

This Agreement constitutes the full and entire understanding and agreement between Contractor and Homeowner with regard to the subjects hereof and supersedes all prior written communications, proposals, understandings, course of dealing, agreements, contracts, and the like between the Contractor and Homeowner.

### Modification and Change Orders

Neither this Agreement nor any term hereof may be amended, waived, discharged, or terminated, except by a writing signed by both the Contractor and Homeowner.

### Jointly Drafted

This Agreement shall be deemed to have been drafted by both parties, and in the event of a dispute, shall not be construed against either party.

Signed this day DAY of MONTH, YEAR.

Contractor: _____

Signed on: _____

Signature: _____

Print name: _____

Address: _____

City, State, Zip: _____

Phone: _____

State license #: _____


Homeowner: _____

Signed on: _____

Signature: _____

Print name: _____

Address: _____

City, State, Zip: _____

Phone: _____


Witnessed: _____

Signed on: _____

Signature: _____

Print name: _____

*Copyright © 2004 Utahbuild.com (www.utahbuild.com). Contract reprinted with permission.*

# Sample Lien Waiver

Property Owners: _____

Owner Address of Record: _____

_____

Property Location: _____

_____

Contractor Name and Address: _____

_____

Description of Work: _____

_____

_____

I/My company was hired by the homeowners listed above to
DESCRIPTION OF ALL WORK TO BE PERFORMED at PROPERTY
ADDRESS. The materials I/my company provided included: ALL
MATERIALS PROVIDED

*Status of Work:*

I state that the work for which I/my company was hired is completed, and I and my subcontractors have been paid in full.
There is no further payment due for materials or labor related to
this project and I waive any right to file a lien against PROPERTY
OWNERS.

Contractor: _____

Signature: _____

Date: _____

Witness: _____

# INDEX

## A

access areas, design process, 77
accessories, bathroom design
  guidelines, 227-228
acrylic-impregnated wood
  flooring, 198
agreements (contracts)
    architects, 116-117
    contractors, 268-270
air conditioning systems,
  177-178
air-to-air heat pumps, 177
angles, cost considerations,
  91-92
appliances, kitchens, 217-219
application process (loans),
  29-30
    documentation needed,
      46-49
      assets, 48
      credit problems, 49
      debts, 48-49
      income, 47
    fees, 41
appraisal fees, 41
arches
    cost considerations, 91-92
    windows, 140
architects
    contract agreements,
      116-117
    costs, 112-113
    floor plan modifications,
      108-109

landscape architects, 249
requesting references from,
  113
selecting, 109-114
working with, 117-119
architectural design plans
    components, 114-116
    contract agreements,
      116-117
    costs, 112-113
    selecting an architect,
      109-114
    working with your archi-
      tect, 117-119
asking prices, land lots, 70
asphalt driveways, 236
asphalt shingles, 148-149
assets
    documentation needed for
      loan application process,
      48
    loan qualification process,
      25
attics, stairs, 185-186
auctions, locating land lots, 61
awning windows, 139

## B

backfilling foundations, 133
balloon framing, 134-135
balloon payment mortgages,
  38
banks, mortgage loans, 31

basements
    cost-cutting tips, 133-134
    foundations, 94
    plumbing, 94
    stairs, 186
    walls, 132
bathrooms
    cost considerations, 90
    cost-cutting tips, 231
    design guidelines
      accessories, 227-228
      general structure, 226
      heat concerns, 228
      shower and tub, 227
      toilets, 227
    fixtures, 230-231
    sinks, 230
    toilets, 229
    tubs and showers, 229-230
bay windows, 139
bedrooms, design process,
  74-75
bids and estimates (contractors), 262
    cost-cutting tips, 264-266
    handling changes, 264
    negotiation process,
      263-266
    suppliers, 274-275
bi-fold doors, 206
bi-level houses, 7
bi-weekly mortgage payments,
  40
blueprints. *See* floor plans

bow windows, 139
brick siding, 144-145
brochures, design ideas, 84
brokers (mortgage), 32
budgets
    challenges, 19
    creating a budget sheet,
        296-298
    dealing with unexpected
        expenses, 310-311
    final cost estimates
        hidden costs, 283
        home value concerns,
            282-283
        reserves, 283, 286
        sample estimate work-
            sheet, 284-285
builders
    design ideas, 83-84
    locating land lots, 61
building process tips, 16
    budget challenges, 19
    costs, 18
    dealing with unexpected
        events, 20
    delays, 19
    doors
        cost-cutting tips,
            143-144
        exterior entry doors,
            142-143
    foundations
        backfilling, 133
        basement walls, 132
        cost-cutting tips,
            133-134
        footings, 130-131
        slabs, 131-132
        waterproofing, 133
    framing
        balloon, 134-135
        cost-cutting tips, 138

floors, 136
        platform, 134
        post-and-beam, 135
        sheathing, 137
        sills, 135-136
        walls, 136-137
    lot preparations, 126-130
    organization, 17
    resources, 17-18
    roofs
        cost-cutting tips,
            152-153
        gutters and downspouts,
            150-151
        soffits and fascias, 152
        steep-slope roof systems,
            148
        types, 148-150
        ventilation concerns,
            151-152
    siding
        brick, 144-145
        cedar shake, 145
        cost-cutting tips, 147
        stone, 146
        stucco, 146-147
        vinyl, 145
        wood clapboard, 146
    skepticism, 20
    time constraints, 17
    versus rebuilding, 14-15
    windows
        cost-cutting tips,
            143-144
        grades, 140-141
        installation problems,
            141-142
        types, 139-140
        U-value, 141
built-in spaces
    cost considerations, 92
    shelving, 182-183

bulk discounts, 312
buyers, finding the right buyer,
    305
bypass doors, 207

## C

cabinets (kitchens)
    construction, 221
    custom, 220
    hardware, 222
    interior fittings, 221-222
    overlay, 221
    semi-custom, 220
    stock cabinets, 219
café doors, 206-207
cape cod style houses, 6
carpet
    cost-cutting tips, 204-205
    pile fibers, 202-204
    styles, 201-202
casement windows, 139
casings (drywall), 196
cast iron pipes, 157
Catalist Homes website, 63
cathedral ceilings, 90
CCA pressure-treated wood,
    244
cedar shake siding, 145
ceilings
    cost considerations, 90
    energy-efficient options,
        98-99
cellulose insulation, 187
ceramic tile flooring, 200-201
checklist, priority, 8
chimneys, framing, 95
circulation lines (water
    heaters), 162
clear wood, 198
clearing lots, 127-128

climate
climate-appropriate options, 81
landscaping concerns, 246
closets, framing, 182
closings
costs, 40-45
documentation, 51-52
quick-closing discounts, 71
code requirements, energy-efficient options, 99
colonial-style houses, 6
commissions, negotiating realty, 62
commitment fee (loans), 41
common wood, 198
composite material decking, 244
concrete driveways, 237
conflicts, contractors, 272-274
construction loans
construction-to-permanent loan, 34
costs, 37
draws and disbursements, 35-36
estimating needed funds, 34
general contractors, 37
land purchase requirements, 35
permit requirements, 36
traditional, 33-34
versus conventional mortgages, 33-34
construction phases
estimating length of, 289
overlapping jobs, 290-293
contract agreements
architects, 116-117
contractors, 268-270

contractors
acting as your own contractor, 256-258
bids and estimates, 262-266
cost-cutting tips, 264-266
handling changes, 264
negotiation process, 263-266
contract agreements, 268-270
custom homes, 3-4
dealing with conflicts, 272-274
communication, 273
legal actions, 273-274
written documentation, 273
finding, 258
fostering good relations with, 271-272
general contractors
construction loans, 37
roles, 253-254
interviewing, 259-262
licensing, 255
payment schedules, 270-271
permits and inspections, 267
production homes, 4
project managers, 255
property liens, 266-267
self-contractors, 4-5
subcontractors, 254
supplying needs, 130
warnings, 258-259
conventional mortgages
balloon payments, 38
fixed and variable rates, 37-38
interest-only loans, 38-39
versus construction loans, 33-34

cooktops, kitchen design guidelines, 213-215
cooling systems, 177-179
cost-cutting tips, 178-179
thermostats, 178
copper pipes, 157
copying floor plans, 105-106
cost considerations
architects, 112-113
basements, 133-134
bathrooms, 231
bids and estimates (contractors), 264-266
budget sheets, 296-298
building process tips, 18
bulk discounts, 312
closing costs, 40, 43, 45
construction loans, 37
cost of property worksheet, 69
dealing with unexpected expenses, 310-311
decks, 245
drywall, 190-191
electrical systems, 173-174
energy-efficient options, 96-99
exterior elements
facades, 88
front porches and entrances, 89
height modifications, 88
odd angles, 87-88
roofs, 89
side-entrance garages, 89
single versus multiple stories, 88
skylights, 89-90
windows and doors, 89
final cost estimate
hidden costs, 283
home value concerns, 282-283

reserves, 283, 286
sample estimate work-
sheet, 284-285
floor plans, 115
build up, not out con-
cept, 120
energy efficiency,
119-120
HUD guidelines, 119
open floor plans, 120
square footage, 121
flooring, 204-205
foundation, 92-94
framing, 94-95, 138
heating and cooling sys-
tems, 178-179
interior elements
arches and angles, 91-92
bathrooms, 90
built-in spaces, 92
doors, 91, 207
electrical needs, 91
fireplaces, 92
open spaces, 90
plumbing, 91
staircases, 91
sunken features, 91
vaulted and cathedral
ceiling, 90
kitchens, 224-225
land lots
cost of property work-
sheet, 69
flood zones, 64
height restrictions, 65-66
negotiating prices, 69-71
special assessments, 64
variances, 64-66
moving, 309
networking, 311
online resources, 312

painting, 193
patios, 246
planning phase, 2-3
plumbing, 166-167
reduction of future costs
(design process), 81
roofs, 152-153
siding, 95, 147
square footage, 77-78
windows and doors,
143-144
counter space, kitchen design
guidelines, 216
countertops (kitchens)
laminate, 222
sinks and fixtures, 223-224
solid composites, 222
stone, 223
tile, 223
wood, 223
crawl space foundations, 93-94
credit
documentation needed for
loan application process,
49
improving, 20-24
credit scores, 22-24
obtaining a credit report,
21-22
credit unions, mortgage loans,
31-32
cross sections, floor plans, 114
crown molding, 195
custom kitchen cabinets, 220
custom homes, general contractors, 3-4
cut-pile carpeting, 201
cuts, wood flooring, 199

**D**

debt
debt-to-income ratio, 26-27
documentation needed for
loan application process,
48-49
loan qualification process,
25
decision-making phase (design
process), 79-81
delaying upgrades, 80-81
dream versus reality, 80
"must have" checklist, 80
why factor, 80
decks
composite material decking,
244
considering factors,
242-243
cost-cutting tips, 245
pressure-treated lumber,
244
vinyl decking, 244
defining your dream home, 5-8
priority checklist, 8
reality check, 6
types of houses, 6-7
delays, building-process tips,
19
delivery basics (suppliers),
277-278
demand water heaters, 161
design process, 73
access areas, 77
bathrooms
accessories, 227-228
general structure, 226
heat concerns, 228
shower and tub, 227
toilets, 227

costs of extra square
footage, 77-78
decision-making phase,
79-81
delaying upgrades, 80-81
dream versus reality, 80
"must have" checklist, 80
why factor, 80
eating and food-preparation
space, 74
factors to consider
neighborhood, 79
slope, 79
view, 78
floor plans. *See* floor plans
free ideas and information,
83-86
kitchens
cooktops and ovens,
213-215
counter space, 216
dishwashers, 215
doorways, 213
food-prep areas, 214
garbage containers, 217
"kitchen triangle," 212
lighting, 217
microwaves, 216
refrigerators, 215
seating clearance,
213-214
walkways, 213
work stations, 213
living space, 74
mistakes, 82-83
reduction of future costs, 81
sleeping areas, 74-75
storage areas, 76-77
work areas, 75-76
digging. *See* excavation
dining areas, 74

discounts
bulk, 312
home shows, 85
realty companies, 63
dishwashers, 215
documentation
closings, 51-52
loan-application process,
46-49
assets, 48
credit problems, 49
debts, 48-49
income, 47
organizing, 279-280
doors
cost considerations, 89-91
exterior entry doors,
142-144
interior doors, 205-208
doorways, kitchen design
guidelines, 213
double-hung windows, 139
downspouts, 150-151
draftsperson, floor plan modifications, 108
drainage-waste-vent system,
156-157
draws and disbursements (construction loans), 35-36
dream homes
building-process tips, 16-20
defining, 5-8
dream versus reality con-
cept, 80
driveways
asphalt, 236
concrete, 237
exit points, 238
gravel, 236-237
paving stones, 237
planning, 237-239
shape, 238

side-entrance garages, 239
slope, 238
width, 238
drywall
casings, 196
cost-cutting tips, 190-191
types, 190
duct systems, energy-efficient
options, 97

**E**

easements, evaluating land lots,
60
eating and food-preparation
spaces, 74
electrical systems, 169
cost considerations, 91
cost-cutting tips, 173-174
external power sources, 172
finding an electrician, 173
finish stage, 173
heating systems, 175-176
lighting, 172
low-voltage wiring, 172-173
outlets, 171
rough-in stage, 173
service boxes, 170-171
switches, 171
wiring, 171
electricians, finding, 173
elevations, floor plans, 115
enamel paints, 192
energy-efficient options
ceilings and fans, 98-99
code requirements, 99
duct systems, 97
floor plans, 119-120
high-performance windows,
98
insulation, 97-98
landscaping, 249-251
tighter construction, 96

water heaters, 162
windows, 141
engineered wood flooring, 198
entrances, cost considerations, 89
entry doors
    cost-cutting tips, 143-144
    types, 142-143
entryways, design process, 77
eRealty website, 63
errors. *See* mistakes
estimates
    budget sheets, 296-298
    contractor bids, 262
        cost-cutting tips, 264-266
        handling changes, 264
        negotiation process, 263-266
        suppliers, 274-275
    dealing with unexpected expenses, 310-311
    final cost estimates
        hidden costs, 283
        home value concerns, 282-283
        reserves, 283, 286
        sample estimate worksheet, 284-285
    funds (construction loans), 34
    schedules
        length of construction phases, 289
        overlapping jobs, 290-293
etiquette, neighbors, 293-294
evaluating land lots
    easements, 60
    historic districts, 59-60
    home base, 57
    homeowners association requirements, 59

    soil quality, 57-58
    utility services available, 59
    water tables, 58-59
excavation (lot preparations), 129-130
exit points, driveways, 238
expenses. *See* cost considerations
exterior projects
    cost considerations, 87-90
    decks, 242-245
    doors, 142-144
    driveways, 236-238
    fences, 251
    landscaping, 246-251
    patios, 245-246
    ramps, 241-242
    stairs, 240-241
    walkways and paths, 240
external power sources, 172

**F**

facades, cost considerations, 88
fans, energy-efficient options, 98-99
fascias, 152
faux finishes, 193
Federal Housing Administration. *See* FHA
fences, 251
FHA (Federal Housing Administration), 39
fiberglass entry doors, 143
fiberglass insulation, 187
finance issues
    housing options, 15
    improving your credit, 20-24
        credit scores, 22-24
        obtaining a credit report, 21-22
    loans. *See* loans

finding
    architects, 109-114
    contractors, 258
    electricians, 173
    land lots, 53-63
    plumbers, 165-166
finishing stage
    electrical systems, 173
    plumbing, 164
fireplaces
    cost considerations, 92
    gas, 208
    wood-burning, 208
fixed rate mortgages, 37-38
fixtures
    bathrooms, 230-231
    kitchens, 223-224
flat doors, 206
flat panel doors, 206
flood certification fee, 41
flood zones, 64
floor plans
    architectural designs, 109-116
    checklists, 123
    cost-cutting tips, 119-121
    creating rough layouts, 101-102
    design process. *See* design process
    hiring a professional to modify plans, 107-109
    home design software, 106-107
    off-the-shelf floor plans, 103-105
    price comparisons, 115
    reproducing, 105-106
    resale value considerations, 122-123
    websites, 104

flooring, 196-205
    carpet, 201-204
    ceramic tile, 200-201
    cost-cutting tips, 204-205
    laminate, 199-200
    vinyl, 197
    wood, 197-199
floors, framing, 136
foam insulation, 188
food-prep areas, design guide-
    lines, 214
footings
    floor plan specifications,
        114
    foundations, 130-131
for sale by owner. *See* FSBO
forced-air heating systems, 176
foundations
    backfilling, 133
    basement walls, 132
    basements, 94
    cost consideration, 92-94,
        133-134
    crawl spaces, 93-94
    excavation, 129-130
    footings, 130-131
    overview, 92-93
    pier foundations, 93
    plan specifications, 114
    slabs, 93, 131-132
    waterproofing, 133
foyers, design process, 77
framing
    balloon, 134-135
    chimneys, 95
    closets, 182
    cost considerations, 94-95,
        138
    floors, 136
    interior walls, 182
    open spaces, 94
    plan, 114

platforms, 134
post-and-beam, 135
room dividers, 182
sheathing, 137
shelving, 182-183
sills, 135-136
trusses, 95
walls, 136-137
free ideas and information
    (design process)
    brochures, 84
    builder's models, 83-84
    home shows, 84-85
    home-related retailers, 85
    landscaping, 248-249
    local colleges, 85-86
    open houses, 85
    websites, 84
french doors, 206
friezé carpets, 202
front porches, cost considera-
    tions, 89
FSBO (for sale by owner),
    locating land lots, 60-61
fuel oil heating systems,
    174-176

**G**

galvanized pipes, 158
garages, side-entrance
    cost considerations, 89
    driveways, 239
garbage containers, kitchen
    design guidelines, 217
gas fireplaces, 208
gas lines, 164-165
general contractors
    construction loans, 37
    custom homes, 3-4
    production homes, 4
    roles, 253-254
geothermal heat pumps, 177

gloss paints, 192
government loan programs
    FHA (Federal Housing
        Administration), 39
    VA (Veteran's
        Administration), 39-40
grades
    windows, 140-141
    wood flooring, 198-199
grading, lot preparations,
    128-129
gravel driveways, 236-237
ground source heat pumps,
    177
gutters, 150-151

**H**

hallways, design process, 77
hardware, cabinets, 222
hardwood flooring
    cuts, 199
    grades, 198-199
    styles, 197-198
heat pump water heaters, 161
heat pumps, 177
heating systems
    cost-cutting tips, 178-179
    fueling sources, 174-176
    thermostats, 178
    types, 176-177
height modifications, cost con-
    siderations, 88
height restrictions, land lots,
    65-66
hidden costs
    final cost estimates, 283
    land lots
        flood zones, 64
        height restrictions, 65-66
        special assessments, 64
        variances, 64-66

historic districts, evaluating land lots, 59-60
home design software, 106-107
home kits, 102-103
home shows
   design ideas, 84-85
   discounts, 85
homeowners associations, 59
homes
   building-process tips, 16-20
   defining your dream home, 5-8
hot water baseboard heating systems, 176
Housing and Urban Development. *See* HUD
housing options, 15
HUD (Housing and Urban Development), 119
HVAC system
   cooling, 177-178
   cost-cutting tips, 178-179
   heating, 174-177
   thermostats, 178

**I**

ideas, design process
   brochures, 84
   builder's models, 83-84
   home shows, 84-85
   home-related retailers, 85
   local colleges, 85-86
   open houses, 85
   websites, 84
improving your credit, 20-24
   credit scores, 22-24
   obtaining a credit report, 21-22
income
   debt-to-income ratio, 26-27
   documentation needed for loan application process, 47

indirect water heaters, 161
inspections
   contractors, 267
   plumbing, 164
insulation, 186
   energy-efficient options, 97-98
   installation, 189
   R-values, 187-189
   types, 187-188
insurance, PMI (private mortgage insurance), 50
interest-only loans, 38-39
interior fittings (cabinets), 221-222
interior projects
   cost considerations, 90-92, 209
   doors, 205-208
   fireplaces, 208
   flooring, 196-205
   framing, 182-183
   insulation, 186-189
   stairways, 183-186
   walls, 189-196
interviewing contractors, 259-262

**J-K**

kitchens, 212
   appliances, 217-219
   cabinets, 219-222
   cost-cutting tips, 224-225
   countertops, 222-224
   design guidelines, 74, 212-217
knobs, interior doors, 207-208

**L**

laminate countertops (kitchens), 222
laminate flooring, 199-200
lanais. *See* patios

land flaws, 70
land lots
   acquiring information at town hall, 66-67
   cost of property worksheet, 69
   evaluating property, 57-60
   hidden costs, 64-66
   locating, 60-63
   negotiating prices, 69-71
   preparations, 126-130
   selection criteria, 53-57
   soil tests, 68-69
   surveying, 68
   water tests, 68
land purchase requirements (construction loans), 35
landings, design process, 77
landscape architects, 249
landscape designers, 249
landscaping
   energy-efficient, 249-251
   free resources, 248-249
   planning, 246-247
   professional versus do-it-yourself, 247-248
latex paints, 192
layouts (floor plans), creating rough layouts, 101-102
lead pipes, 158
legal actions against contractors, 273-274
lenders
   closings, 51-52
   documentation needed for loan application process, 46-49
   inspection fee, 41
   PMI (private mortgage insurance), 50
   secrets, 45-46
LendingTree website, 63
level-loop pile carpets, 202

licensed contractors, 255
liens (property), contractors, 266-267
lighting
  electrical systems, 172
  kitchen design guidelines, 217
living space, design process, 74
loan-origination fees, 40
loans
  application phases, 29-30
  bi-weekly payments, 40
  closings
    costs, 40-45
    documentation, 51-52
  construction, 33-37
  conventional mortgages, 37-39
  documentation needed for application process, 46-49
  finding the right lender, 32-33
  government loan programs, 39-40
  institution options, 31-32
  lender secrets, 45-46
  PMI (private mortgage insurance), 50
  prequalification, 30-31
  qualification phase, 24-27
locks, interior doors, 207-208
long-wearing options, 81
lots. See land lots
low-voltage wiring, 172-173

M

maintenance-free options, 81
management (project), schedules, 295
materials
  suppliers, 274-277
  take-off lists, 280-282
  theft, 278

measurements, stairways
  total rise, 183-184
  total run, 183-184
metal shingles, 150
microwaves, kitchen design guidelines, 216
mistakes, design process, 82-83
model homes, design ideas, 83-84
modern homes, 7
modifying floor plans, 104-105
modular home kits, 102-103
molding
  crown, 195
  types, 196
mortgage brokers, 32
  fees, 41
mortgages
  application phases, 29-30
  bi-weekly payments, 40
  closing documentation, 51-52
  conventional mortgages
    balloon payments, 38
    fixed and variable rates, 37-38
    interest-only loans, 38-39
    versus construction loans, 33-34
  documentation needed for application process, 46-49
  finding the right lender, 32-33
  government loan programs, 39-40
  institution options
    banks, 31
    credit unions, 31-32
    mortgage brokers, 32
  lender secrets, 45-46
  prequalification, 30-31
  PMI (private mortgage insurance), 50

moving tips
  cost-cutting, 309
  housing options, 15
  moving companies, 306
  planning, 306-309
  releasing your rental, 299-300
  selling your home, 300-305
multilevel loop pile carpets, 202
"must have" checklist (design process), 80

N

National Roofing Contractors Association. See NRCA
natural gas heating systems, 175
negotiations
  bids and estimates (contractors), 263-266
  land lot prices
    area lot comparisons, 69
    asking prices, 70
    land flaws, 70
    quick-closing discounts, 71
  realty commissions, 62
neighborhood, design process considerations, 79
neighbors, etiquette recommendations, 293-294
networking, 311
NRCA (National Roofing Contractors Association), 148

O

off-the-shelf floor plans
  modifications, 104-105
  sources, 103-104

oil paints, 192
online resources, cost-cutting
 tips, 312
open floor plans, 120
open spaces
 cost considerations, 90
 framing, 94
organization
 building-process tips, 17
 documentation, 279-280
outlets, electrical systems, 171
ovens
 design guidelines, 214
 kitchen design, 213-215
overlay, cabinets, 221

**P**

painting
 cost-cutting tips, 193
 faux finishes, 193
 types of paints, 192
 walls, 191-193
paneling walls, 191
parquet flooring, 198
pathways, 240
patios, 245-246
paving stone driveways, 237
payment schedules, contractors, 270-271
permits
 construction loan requirements, 36
 contractors, 267
 plumbing, 163
pier foundations, 93
pile fibers (carpets), 202-204
pile foundations. *See* pier foundations
pipes (plumbing systems)
 cast iron, 157
 copper, 157
 galvanized, 158

lead, 158
plastic, 158
sizes, 159
plainsawn wood floors, 199
plank flooring, 198
planning
 cost-cutting tips, 2-3
 driveways, 237-239
 landscaping, 246-247
 moving tips, 306-309
 plumbing, 163-164
plans. *See* floor plans
plaster walls, 191
plastic pipes, 158
platform framing, 134
plot plans, 115
plumbers, finding, 165-166
plumbing systems, 155
 basements, 94
 cost considerations, 91,
 166-167
 drainage-waste-vent system,
 156-157
 finding a plumber, 165-166
 inspections, 164
 pipes, 157-159
 planning, 163-164
 shutoff valves, 159
 supply systems, 156
 water heaters, 160-162
plush/velvet carpet, 201
PMI (private mortgage insurance), 50
pocket doors, 207
post-and-beam framing, 135
prequalification (loans), 30-31
pressure-treated lumber
 (decks), 244
priority checklist, 8
private mortgage insurance.
 *See* PMI
production home builders, 4

professionals
 design, 107-114
 architects, 108-114
 draftsperson, 108
 landscaping, 247-248
project management
 budgets, 296-298
 schedules, 295
project managers, roles, 255
propane heating systems, 175
property liens, 266-267

**Q**

qualifying for loans, 24-27
 assets, 25
 debt concerns, 25
 debt-to-income ratio, 26-27
quartersawn wood floors, 199
quick-closing discounts (land
 lots), 71

**R**

radiant heat, 177
railings (stairways), 184-185
raised panel doors, 206
ramps, 241-242
ranch-style houses, 6
real estate agents
 discount realty companies,
 63
 locating land lots, 61-63
 negotiating commission, 62
 selling your home, 300-302
realty commissions, negotiating, 62
rebuilding versus building,
 14-15
reduction of future costs
 (design process), 81
references, architects, 113
reflective material insulation,
 188

refrigerators, kitchen-design guidelines, 215

rentals, moving tips, 299-300

reproducing floor plans, 105-106

resale value considerations, 122-123

reserves, final cost estimate, 283, 286

resources
building-process tips, 17-18
landscaping, 248-249
product offerings, 15-16

retailers, design ideas, 85

riftsawn wood floors, 199

rock and slag wool insulation, 188

role of contractors
general contractors, 253-254
project managers, 255
subcontractors, 254

roofs
cost considerations, 89, 152-153
gutters and downspouts, 150-151
soffits and fascias, 152
steep-slope roof systems, 148
types, 148-150
ventilation concerns, 151-152

room dividers, framing, 182

rough-in stage
electrical systems, 173
plumbing, 163-164

R-values (insulation), 187-189

**S**

samples
estimate worksheets, 284-285
schedules, 290-293

saxony carpets, 202

schedules
estimating length of construction phases, 289
overlapping jobs, 290-293
planning for delays, 286-289
project management, 295

seating clearance, kitchen design guidelines, 213-214

select woods, 198

self-contractors, 4-5

selling your home, 300
finding the right buyer, 305
moving tips, 305
cost-cutting tips, 309
moving companies, 306
planning, 306, 309
real estate agents, 300, 302
setting the price, 304-305
timing, 303-304
tips to spruce up your home, 302-303

semi-custom cabinets, kitchens, 220

septic systems, 168-169

service boxes (electrical systems), 170-171

shade, energy-efficient landscaping, 250

shape of driveways, 238

sheathing, 137

shelving, framing, 182-183

shingles
asphalt, 148-149
metal, 150
slate, 150

synthetic, 150
tile, 149
wood, 149

showers
bathroom design guidelines, 227
styles, 229-230

shutoff valves (plumbing systems), 159

side-entrance garages
cost considerations, 89
driveways, 239

siding
brick, 144-145
cedar shake, 145
cost considerations, 95, 147
stone, 146
stucco, 146-147
vinyl, 145
wood clapboard, 146

sills (framing), 135-136

single-story houses, cost considerations, 88

sinks
bathrooms, 230
kitchens, 223-224

sizes
plumbing pipes, 159
water heaters, 160

skylights, cost considerations, 89-90

slab foundations, 93, 131-132

slate roofing products, 150

sleeping areas, design process, 74-75

slider windows, 140

sliding doors, 207

slope
design-process considerations, 79
driveways, 238
steep-slope roof systems, 148

soffits, 152
software, home-design software, 106-107
soil
    evaluating land lots, 57-58
    landscaping, 247
    lot preparations, 128-129
    testing, 68-69
solar heating systems, 175
solar water heaters, 162
solid composite countertops (kitchens), 222
solid hinged doors, 205-206
solid wood flooring, 197
special assessments (land lots), 64
specifications, floor plans, 114
split-level homes, 7
square footage, cost considerations, 77-78, 121
stairways
    attics, 185-186
    basements, 186
    cost considerations, 91
    exterior, 240-241
    railings, 184-185
    total rise measurement, 183-184
    total run measurement, 183-184
    width, 184
steel entry doors, 142
steep-slope roof systems, 148-150
stock cabinets (kitchens), 219
stone countertops, 223
stone siding, 146
storage areas, design process, 76-77
storage water heaters, 160-161
stress-management techniques, 310

strip flooring, 198
stucco siding, 146-147
subcontractors, roles, 254
sunken features, cost considerations, 91
Superfund sites, 57
suppliers
    bids and estimates, 274-275
    delivery basics, 277-278
    questions to ask, 275, 277
    take-off lists, 274-275
supply system (plumbing), 156
surveying land lots, 68
switches, electrical systems, 171
synthetic roofing products, 150

**T**

take-off list, 274-275, 280-282
tankless coil water heaters, 161
testing land lots
    soil test, 68-69
    surveying, 68
    water test, 68
theft of building materials, 278
thermostats, 178
tile
    countertops, 223
    flooring, 200-201
    roofing products, 149
time constraints, 17
timing the sale of your home, 303-304
toilets
    bathroom-design guidelines, 227
    styles, 229
total rise measurement (stairways), 183-184
total run measurement (stairways), 183-184
tracking pads (lot preparations), 129

traditional construction loans, 33-34
transom windows, 140
trim
    crown molding, 195
    types, 196
trusses, framing, 95
tubs
    bathroom design guidelines, 227
    styles, 229-230
two-story houses, cost considerations, 88

**U**

underwriting fees, 41
upgrades, delaying, 80-81
usage planning, 181
utilities
    energy-efficient options, 96-99
    evaluating land lots, 59
utility rooms, 231-232
U-value (windows), 141

**V**

VA (Veterans Administration) loans, 39-40
variable rate mortgages, 37-38
variances, 64-66
vaulted ceilings, cost considerations, 90
ventilation
    roofs, 151-152
    systems, 178
view, design process considerations, 78
vinyl
    decking, 244
    flooring, 197
    siding, 145

# W

wainscoting, 195
walkways
    design guidelines, 213
    exterior, 240
wallpapering, 194-195
walls, 189
    basement, 132
    drywall
        cost-cutting tips,
            190-191
        types, 190
    framing, 136-137, 182
    molding, 195-196
    painting, 191-193
        cost-cutting tips, 193
        faux finishes, 193
        types of paints, 192
    paneling, 191
    plaster, 191
    wainscoting, 195
    wallpapering, 194-195
warnings, contractors, 258-259
water conservation, energy-
    efficient landscaping,
    250-251
water heaters
    circulation lines, 162
    demand water heaters, 161
    energy efficiency, 162
    heat pump, 161
    indirect, 161
    location concerns, 160
    sizes, 160
    solar, 162
    storage, 160-161
    tankless coils, 161
water supply
    plumbing. *See* plumbing
        systems

wells
    digging, 167-168
    factors to consider,
        167-168
water tables, evaluating land
    lots, 58-59
water tests, 68
waterproofing foundations,
    133
websites
    Catalist Homes, 63
    design ideas, 84
    eRealty, 63
    floor plans, 104
    LendingTree, 63
    ZipRealty, 63
wells
    considering factors,
        167-168
    digging, 167-168
why factor (design process), 80
width
    driveways, 238
    stairways, 184
wind reduction, energy-
    efficient landscaping, 250
windows
    cost considerations, 89,
        143-144
    grades, 140-141
    high-performance, 98
    installation problems,
        141-142
    types, 139-140
    U-value, 141
wire transfer fees, 41
wiring
    electrical systems, 171
    low-voltage, 172-173

wood
    pressure-treated lumber
        (decks), 244
    clapboard siding, 146
    countertops (kitchens), 223
    entry doors, 142
    flooring
        cuts, 199
        grades, 198-199
        styles, 197-198
    shingles, 149
    wood-burning fireplaces,
        208
work stations
    design process, 75-76
    kitchen-design guidelines,
        213
worksheet, estimate, 284-285
worth, home value concerns,
    282-283

# X-Y-Z

ZipRealty website, 63